CAMBRIDGE LIBRARY C

Books of enduring scholarly value

History

The books reissued in this series include accounts of historical events and movements by eye-witnesses and contemporaries, as well as landmark studies that assembled significant source materials or developed new historiographical methods. The series includes work in social, political and military history on a wide range of periods and regions, giving modern scholars ready access to influential publications of the past.

How to Develop Productive Industry in India and the East

The craft of manufacturing cotton was learnt by the British in India, but the East India Company's drive for profits caused the development of a cotton-spinning industry in Britain, which grew enormously when mechanical processes were introduced during the industrial revolution. Subsequent protectionist legislation which forbade the export of anything except raw cotton meant that India's supremacy in the industry was lost to Britain in the nineteenth century. This 1867 work, edited by P.R. Cola, who owned the Arkwright Cotton Mills in Bombay (Mumbai), argues for investment in up-to-date machinery, and provides a blueprint for developing a host of industries such as cotton and other textiles, jute, sugar, oil and iron, which would bring prosperity to India. Containing illustrations and statistical data, the book gives useful insights into the early development of many industries in India and in some of the other economies of the East.

Cambridge University Press has long been a pioneer in the reissuing of out-of-print titles from its own backlist, producing digital reprints of books that are still sought after by scholars and students but could not be reprinted economically using traditional technology. The Cambridge Library Collection extends this activity to a wider range of books which are still of importance to researchers and professionals, either for the source material they contain, or as landmarks in the history of their academic discipline.

Drawing from the world-renowned collections in the Cambridge University Library and other partner libraries, and guided by the advice of experts in each subject area, Cambridge University Press is using state-of-the-art scanning machines in its own Printing House to capture the content of each book selected for inclusion. The files are processed to give a consistently clear, crisp image, and the books finished to the high quality standard for which the Press is recognised around the world. The latest print-on-demand technology ensures that the books will remain available indefinitely, and that orders for single or multiple copies can quickly be supplied.

The Cambridge Library Collection brings back to life books of enduring scholarly value (including out-of-copyright works originally issued by other publishers) across a wide range of disciplines in the humanities and social sciences and in science and technology.

How to Develop Productive Industry in India and the East

Mills and Factories

P.R. Cola

CAMBRIDGE
UNIVERSITY PRESS

CAMBRIDGE UNIVERSITY PRESS

Cambridge, New York, Melbourne, Madrid, Cape Town,
Singapore, São Paolo, Delhi, Mexico City

Published in the United States of America by Cambridge University Press, New York

www.cambridge.org
Information on this title: www.cambridge.org/9781108046237

© in this compilation Cambridge University Press 2012

This edition first published 1867
This digitally printed version 2012

ISBN 978-1-108-04623-7 Paperback

HOW TO DEVELOPE

PRODUCTIVE INDUSTRY IN INDIA

AND THE EAST.

MILLS AND FACTORIES

FOR

GINNING, SPINNING, AND WEAVING COTTON; JUTE AND SILK MANUFACTURES;
BLEACHING, DYEING, AND CALICO PRINTING WORKS; SUGAR, PAPER,
OIL, AND OIL-GAS MANUFACTURES; IRON AND TIMBER WORK-
SHOPS; CORN-MILLS, ETC., ETC.

WITH

ESTIMATES AND PLANS OF FACTORIES.

EDITED BY P. R. COLA,

LATE SOLE PROPRIETOR OF THE ARKWRIGHT COTTON MILLS, BOMBAY.

Illustrated with more than One Hundred Woodcuts.

LONDON: VIRTUE AND CO.

CALCUTTA: LEPAGE & CO. BOMBAY: CHESSON & WOODHALL.
MADRAS: G. HIGGINBOTHAM.
LAHORE: W. COCKELL & CO. AGRA: DELHI GAZETTE PRESS.
MOOLTAN: DADABHOY MANEKJEE.

1867.

LONDON:
PRINTED BY VIRTUE AND CO.,
CITY ROAD.

PREFACE.

At a time when the Metropolis of India is exchanging, in the course of a few hours, telegraphic messages with the Metropolis of England—when the capital of China will also be shortly brought within the electric circuit—when the West and the East are united with each other in friendly intercourse—when the English, French, and Indian steamers are carrying mails and merchandise every week backwards and forwards—in fact, when the civilised West has been brought into such close proximity to the East—

it is of the highest importance that the improved machinery of modern times should be substituted in place of the rude apparatus with which the arts and manufactures are still carried on in India and the East. As one means of furthering so desirable an object, this volume is published; but not with the idea that a country is necessarily richer by manufacturing its own raw produce. Let this be done by any nation, it matters not in what part of the world, as long as it can be done, and done well, and brought to the market at the cheapest price.

Plans of factories are given for some of the most important industrial works now carried on in the East with rude appliances, but which, if worked on the factory system with improved machinery, there is every reason to believe, will prove commercially successful.

The fact that capitalists in India and the East have not that enterprise and spirit which distinguish the European race, has been kept in mind in fixing the size of works. In the East, persons with large fortunes, and generally not educated, have naturally no inducements to trouble their minds with manufacturing enterprise; they therefore require to be introduced to the subject by degrees. Estimates are given and brought to such a point that persons possessing moderate capital will be able to carry out a great many of the industrial projects described. This will be preferable to the formation of joint-stock companies for that purpose. But joint-stock companies, if well managed, will be able to erect works on a scale of magnitude which would vie with those of Europe, where single individual firms invest such large amounts of capital as could hardly be conceived by persons possessing the same wealth in the East.

Attention has been given, in the plans of mills and factories, to adapt them to the wants of the country. The estimates have been framed in several cases from actual invoices; and the practical information has been gathered in the manufacturing districts of England from the best sources. The introductory and manu-facturing chapters to each subject have been compiled from very reliable authorities, and the whole put into such shape, and so illustrated, as to prove a useful industrial guide-book, not only in India and the East, but also in other countries. Even to the general reader, not connected with factories or mills, but taking an interest in the advancement of civilisation and the elevation of mankind, the introductory chapters, in which the rise, progress, and present state of each manufacture are given, will, it is hoped, be of some interest. The statistics from the papers presented to Parliament, and other authentic sources, will be of use as showing the direction in which industrial progress can be developed, and to what extent it is capable of expansion.

Though this work is published in the interest of India and the East, it will not be the less in the

interest of England and the English machinists, as
what will benefit the one country must benefit the
other also.

To name those firms and individuals who have given
valuable information would be to fill up one or two
pages with mere names. To name one and not the
others would be invidious. To one and all, therefore,
the Editor tenders his warmest and most sincere
thanks. May the Almighty promote the interests of
all countries, in the cause of Peace and Prosperity!

P. R. C.

London, *July*, 1867.

CONTENTS.

LIST OF PLANS OF FACTORIES.

LIST OF ILLUSTRATIONS.

INDUSTRIAL DEVELOPMENT IN INDIA
AND THE EAST.

A SHORT stay in the manufacturing districts of England
—in the midst of huge spinning and weaving mills,
bleach-works, print-works, engine factories, workshops,
and forges, the incessant noise of spindles and machinery, and the endless traffic—brings vividly to the
mind of even a very dull person the advantages which

the use of machinery confers on mankind.   These advantages arise from the vast addition which machinery makes to human power, the economy which it effects in time, the precision with which all its operations are conducted, and the exact similarity in the articles that are made.   It is by the aid of machinery only that the productive work of a single individual can be multiplied a thousand-fold, and made far to surpass in cheapness any hand-made work, notwithstanding that, in the first instance, a great outlay is required in machinery and other stock.

In India and other parts of Asia, as labour is getting scarcer every year, the introduction of improved machinery in place of the rude implements still in use would be beneficial, seeing that economy must arise from the substitution of machines.

It cannot be denied that the elements of manufacturing success exist in India and other parts of Asia; nor that the natives will acquire more wealth by manufacturing the raw material with improved machinery for their own wants and for export, than by sending such materials thousands of miles, entailing a heavy expense in the cost of transit—to be returned in the manufactured state to the very places whence the raw materials were sent in the first instance, and charged with all the immense attendant outlay, which amounts to from twenty-five to thirty per cent.

No doubt the first cost of machinery in India and China will be higher than in Great Britain, arising from the freight and other charges which would have to be incurred; and much more will have to be paid for fuel; but, on the other hand, the saving in the price of raw material will more than equal all the transit, insurance, and merchants' charges. To take an illustration. Despite the drawback in Bombay of having to pay for coals much more than is paid in Great Britain—with the disadvantage of untrained hands in the mills, producing a less quantity of yarn and cloth than in England per a given number of spindles and looms, and bad management in nine mills out of ten— the cost of spinning No. 20's twist (the principal number in demand in the Bombay Presidency) does not come to more than the cost of spinning the same number in a Lancashire mill in England. The high price of fuel, moreover, is in course of being lessened by the working of coal mines. The East India Railway Company obtains its supplies from the mines of Burdwan Raneegunge at a cost of no more than ten shillings per ton at the mouth of the pit.

That improved machinery can be successfully employed in India and China for particular manufactures, and more cheaply than hand labour, cannot be doubted. Its success has been proved in the case of cotton mills in Bombay, and jute mills near Calcutta—the latter of which are almost all owned by Europeans.

It has been asserted that India is only an agricultural country; but it is an agricultural and a manufacturing country as well. No better proof of this could be given than the seven hundred working samples of Indian textile fabrics, collected in eighteen volumes, and presented by the Secretary of State for India to the Manchester Chamber of Commerce and other institutions. The commissioner sent by the Manchester Cotton Supply Association to India, during the American war, stated in his official report that 1,250,000,000 pounds of cotton are consumed in India, and that 33,000,000 native spindles must be kept at work by the people of India.

The birth-place of cotton manufactures was India. India supplied Great Britain with yarn and cotton goods long before she furnished a pound of the raw material. Jaconets, mulmuls, doreas, were all originally manufactured in India. India not only carried on internal manufacture, but also an export trade. The Europeans carried the knowledge of this manufacture originally from India to Europe. The general name of Calico has been applied to the plain white cloth, from the circumstance of this article having been first exported from Calicut, situated on the Malabar Coast, in India. In the seventeenth century, the Dutch and English East India Companies sent to Europe Indian manufactured goods in large quantities—the finest muslins and table-cloths from Bengal; best chintzes, ginghams, and long cloths from Madras, and strong and inferior goods from Surat and Calicut. For many hundred years Persia, Arabia, and all the eastern parts of Africa were supplied with a considerable portion of their cloths, cottons, and muslins, from the marts of India.

In 1621, Mr. Main, one of the Directors of the East India Company, estimated the annual importation of Indian calicoes in Great Britain at 50,000 pieces, the average cost of which on board in India was seven shillings. The Indian muslins, chintzes, and calicoes became very fashionable in Britain for ladies' and children's dresses. To such an extent did this proceed, that as early as 1678 a loud outcry was made in England against the admission of Indian goods, which, it was maintained, were ruining the English woollen manufacture—at that time the most extensive branch of manufacture in Great Britain. An Act was passed in the year 1700 (Act 11 and 12, William III., chap. 10) which forbade the introduction of Indian goods for domestic use in Great Britain under a penalty of £5 on the wearer and £20 on the seller. The other governments of Europe also thought it necessary to prohibit the introduction of Indian manufactured goods, or to load them with very heavy duties. In Britain the Act did not prevent the continued use of the goods, so a duty was levied on Indian manufactured textiles, —ten per cent. on cotton, and twenty per cent. on silk goods. That useful fabric known as nankeen was also imported into Great Britain from China directly and indirectly to a large extent.

There can be no doubt that Indian calicoes, muslins, chintzes, and China nankeen had become common in England at the close of the seventeenth century. De Foe, in a number of the *Weekly Review*, January, 1708, sharing the general notion of that time, bitterly lamented the large importations of Indian goods into Great Britain, and stated that " the general fansie of the people runs upon East India goods. . . . It has crept into our houses, bed-chambers, curtains, cushions, chairs ; and at last beds themselves were nothing but calicoes or Indian stuffs ; and, in short, almost everything relating to the dress of the women or the furniture of our houses was supplied by the Indian trade. About half of the woollen manufacture was entirely lost, half of the people ruined, and all this by the intercourse of the East India trade." (!)

But a revolution has since taken place. The tide of commerce now runs more rapidly against the

Indians than it ever ran against the English. And why ? Not because the British have made more progress in the delicacy of manipulation,—even with the aid of machinery the beautiful Dacca muslins have scarcely been surpassed by them as yet,—but simply in the *immense quantity* of goods they can manufacture by the substitution of improved machinery.

The art of printing calicoes by hand blocks was practised in India from the earliest ages : but now the calico printing machinery of Great Britain and other European nations is fast superseding the printing of Italian chintzes by hand labour—though in the year 1680 the East India House was mobbed in revenge for some large importations made by the East India Company, of chintzes manufactured in India.

The process of dyeing Turkey red was also first discovered and practised in India; it was then introduced into Europe and greatly improved, and now a considerable portion of the Turkey-dyed British goods finds its way back again to the East, though the beautiful dyes of India still maintain their high character in Europe. The most important dye-stuffs so extensively used in Great Britain, such as madder, indigo, lac dye, &c., are still produced in the East, and exported in large quantities to Europe.

There is no doubt that, had it not been for the invention and improvement of machinery in Europe, India would still have maintained her supremacy. But while every country in Europe was making some addition to its knowledge in useful arts and sciences, Asia slumbered in a death-like torpor, spell-bound and entranced by the accursed superstition which preyed on her strength.

But now the time is come when Asia must rise from her inactivity, cast off her superstitions, march with the progress and civilisation of the times, and amongst other things (instead of spinning and weaving, dyeing and calico printing, and manufacturing, with rude apparata, just as her forefathers did a thousand years ago) substitute the improved machinery of modern times. India must improve the resources which nature has provided; she must develop industrial appliances, and give facility to the accumulation of wealth by well-directed capital and labour, and by applying improved machinery and the useful arts for the production of various commodities. There is no doubt that the marvellous prosperity of Great Britain has, to a great extent, resulted from her manufactures, chiefly cotton, and that her wealth has been accumulated chiefly by manufacturing enterprise. The condition of any country must be precarious as long as its people are chiefly dependent on the produce of the land. It is the prevalence of manufacturing industry that has placed European countries in a position of prosperity.

A stationary position has ruined Indian manufactures. But in the nineteenth century, under the British Raj, with a weekly communication to and from the most civilised nations on earth, a stationary position is out of the question. The accumulated result of many centuries' invention now lies at the door of Asia. Progress in manufacturing and agricultural industry must henceforth be her motto. It will enrich the people, impart knowledge, remove prejudice and superstition from the land, and enable Asia to share with other nations advantages she does not enjoy at present.

The prosperity of Asia will not depend solely on the

extent of her population nor the richness of her undeveloped natural resources, nor even on her accumulated wealth, but mainly on that portion of her wealth
which is being directed to useful purposes.   It is only
that portion of her capital which may be thus employed
in Asia that will affect her industrial progress.   The
industrial condition of Asia will never be improved by
locking up her wealth in unproductive, useless works,
or by hoarding up the precious metals—a practice
which still prevails to a very great extent.   It will be
mainly by employing her wealth in aiding labour in
the development of her resources, and in the introduction of improved machinery, that her prosperity can
be stimulated.   In a country where the ingredients of
manufacturing success do exist, why should its power
be wasted, instead of employing that power for its own
benefit and the benefit of its teeming millions ?

COTTON GINNING.

INTRODUCTION.

AMONG all the various materials which the skill of man converts into comfortable and elegant clothing, the most extensively useful is the beautiful produce of the cotton plant. It possesses downy softness and warmth, and its delicate fibres are flexible and tenacious. Dr. Forbes Watson, reporter on the products of India, has stated that there is no country in the world of equal extent so favoured as Hindostan for the production of cotton, as regards climate and soil. More than seventy

samples of soil, carefully selected in India, were ana-
lysed in England, and conclusively showed the richness
of the Indian soil in all the important constituents
required for the growth of cotton and other fibrous
plants.

In almost every part of India cotton is grown for
local consumption and for export; but certain districts
are wholly occupied with its cultivation, chiefly for
export to Great Britain and China. The Bombay
Presidency, Surat, Broach, and other sea-indented
lands of Guzerat, with their numerous outlets, have
been the chief sources of supply. In the interior,
Berar, which is celebrated for its superior cotton, has
been able to supply places from 400 to 1,000 miles
distant, finding it sufficiently remunerative. In the
south, another important and extensive cotton district
is in the Southern Mahratta country. To the east of
Berar is the extensive province of Nagpore, where
Oomravathy and Khangaum are the chief cotton marts.
In the Madras Presidency, including Bellary, Tinne-
velly, and other collectorates, thousands of acres are
planted with cotton.

The culture of American cotton has taken firm root
in India, chiefly in Dharwar; and from that district
cotton is now obtained of a quality nearly equal to
middling American, and which on the whole yields
satisfactory profit to the cultivators. For this boon,
a debt of gratitude is due to Mr. Shaw, the collector,
and Dr. Forbes, who caused the average yield of
cotton and its market worth per pound to be nearly
doubled.

The Dharwar cotton, from the acclimatised American
seed, now stands at the head of the Indian price list,

and, according to the report of the Commissioner sent out by the Manchester Cotton Supply Association, is, when properly cleaned, a fair rival to New Orleans cotton.

Dr. Forbes has recently published a report to show what other provinces besides Dharwar are best adapted for the cultivation of the acclimatised American seed; and it appears that the acclimatised plant will thrive in many districts: in the Nizam's territory, a hundred miles from Dharwar—in Bellary in the Madras Presidency—and also in the North-West Provinces. In Khandeish, where cotton was grown of very inferior quality, the defect, according to official statements, has been remedied by the extirpation of the old cotton, and the complete introduction of the Berar variety; and there is every prospect of a grand future for the province, with its large area suitable for the crop.

In the official reports, 1866-7, of the cotton crops of the Southern division of the Bombay Presidency, it was shown that while the total cost to the cultivator per acre, in the case of exotic cotton, was 15s., and for the native plant, 13s. 6d., the profit by the sale of the produce in the former case was 25s., and in the latter, 12s. 6d., showing an advantage in favour of acclimatised New Orleans of 12s. 6d. *per acre*. The extra out-turn not only represents a very much larger profit, but, what is of still more importance in a country where almost every acre of arable land is taken up, it is an immense saving of cultivatable area.

The Manchester Cotton Supply Association has rendered good service to India and other cotton-growing countries by furnishing the best description of seeds and improved machinery; and has issued several pamphlets describing the best modes of culture, for the use of persons engaged in cotton cultivation. The Association looks to India as the country from which the largest supply, next to America, is to be obtained.

There is ample room for the efforts of private enterprise and public companies in various parts of Asia, for the introduction not only of improved seeds, but also of improved agricultural and ginning machinery ; seeing that the cultivators or ryots are a poor and ignorant race, having no money to lay out on the improvement of their land, or in improved machinery for cleaning cotton.

COTTON GINNING FACTORY.

PRACTICAL PROCESSES.

When cotton is plucked from the plant, its fibres are largely mixed with the seeds, and with other woody fibrous matters which adhere very closely. These impurities must be removed before any attempt is made to manufacture it. Considerable difference is found to exist between the different varieties of cotton in the force with which it adheres to the seed. In the black-seeded varieties it separates easily, while in the green-seeded cotton it adheres firmly.

The first process, therefore, is the separation of the seed from the staple to which it is firmly attached, converting seed cotton into clean cotton by means of *gins*. The oldest of these machines are the Indian *churkas*. They are still used in India, and have probably remained without material improvement or modification for two thousand years. In the Indian churka two small wooden rollers are placed one over the other in a frame, and they work together with teeth cut out of the ends. The machine, though it hardly deserves that name, is turned by one person and fed with cotton by another. The kapas, or seed-cotton, in its dirty state, is put in at one side and the fibres drawn through between the rollers; the seeds being too large to pass are thereby torn off, and fall down to the opposite side, being thus separated from the clean fibre. More than 75 per cent. of the cotton produced in India is ginned by the common churka, and the yield of

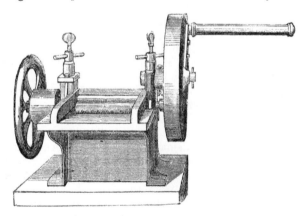

clean fibre is half a pound per hour for each person. The churka used in the Southern Mahratta country is the foot-roller, worked by a woman, which turns out from 8 to 12 lbs. of clean cotton daily. The common Guzerat churka turns out 12 to 14 lbs. of clean cotton and 28 lbs. seeds in a day, and requires two persons to work it. Some years ago Dr. Forbes tried to improve the Indian churka for cleaning kapas or seed-cotton, by introducing improvements of a simple character, to enable one person instead of two to work it.

From the records of the Bombay Chamber of Commerce it appears, according to the minute of Mr. John Flemming (who has been honoured with the order of the "Star of India," and who has done more than any other native or European merchant in introducing useful industrial enterprise into Western India), that Dr. Forbes' improved cottage churka turns out a larger quantity of clean cotton in a given time than the rude Indian apparatus, without injuring the staple.

There is a general complaint that the cotton crop is not properly weeded, and that when ripe it is not carefully picked or properly cleaned ; this arises from the scarcity of hands, which is very great in the picking season. One mode, as Dr. Forbes remarks, of remedying these defects will be the introduction of machinery for cleaning cotton. Machinery will not only ensure that the cotton shall be better cleaned and packed, but will also set free many of the hands now employed on the clumsy and inefficient native churkas.

In America, Mr. Whitney invented his well-known saw-gin, which enabled cotton-growers to keep pace in cleaning it with the demand of the times. An immense stimulus was thus given to the cotton-producing power of America. The production in the United States alone rose within ten years from the date of invention two hundred fold ; and in 1860 the cotton crop in that country amounted to 3,800,000 bales. The cotton gins now in general use in the States are, in principle, the same as the original Whitney gin, though a good many improvements have been introduced from time to time. The gin contains circular saws mounted on a cylinder turned by a fly-wheel. Another cylinder mounted with brushes, and working in an opposite direction, cleans the cotton from the teeth of the saws. It is stated that one man with a two-horse power engine can clean, with a Whitney's eighty-saw gin, about 80 pounds per hour. But the American saw-gin is not adapted for ginning the short-staple variety of Indian cotton, because the fibre is too weak to stand the action of the saws. In the case of the New Orleans cotton of the Dharwar and other districts, a modification of the saw-gin, adopted in the Dharwar Government gin factory, has answered to some extent. These saw-gins are worked by manual labour ; eight persons being employed to a gin of eighteen saws, and six to one of ten saws.

It is strange that at a time when the establishment of railways in India is rapidly extending, and when steam has been applied in England even for such purposes as driving mock wooden horses round and round at a rapid speed for the mere amusement of children (as may be seen at Knott Mill and other fairs in England), and even in brushing hair at hair-cutters' shops, some of the Indian Government officials have seriously proposed *to economise labour and time* by devising a simple gin to be entrusted to common villagers. This is surely a mistake. Economy of labour and celerity of performance will only be insured by the use of cotton-gins provided with the latest improvements, and driven by steam, as lately introduced in Guzerat.

Since the establishment of the first steam cotton-ginning factory, the increased cultivation of cotton in other parts of the world has so increased the demand for cotton-gins, that English machinists have paid within the last few years special attention to their improvement and construction, so that every description of cotton may be ginned by them. The gin introduced into Guzerat is that known as the *Improved Macarthy Gin*, made forty inches wide. The seed cotton is fed

into this machine by an endless travelling apron, opened
out by a spiked roller, transferred therefrom in detached
tufts by the vibratory movements of a transferring comb,
and presented to another roller between fixed and

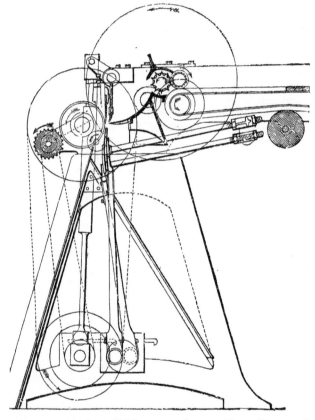

vibrating blades, which come into action alternately.
The seeds are separated in this improved gin without
being crushed, and the fibre of cotton is not in the

least injured. Instead of the leather on the roller in this machine, which requires to be renewed at short intervals of time, and which is expensive, jute-cloth, which is very cheap, has been substituted, to effect economy in working the gin.

It is a gratifying fact that within the last three years about 2,000 Macarthy gins, *worked by steam,* have been introduced into Guzerat, chiefly through the enterprise of a few Europeans; while in Dharwar, producing a far better quality of cotton, and therefore with the sure prospect of obtaining an increased production in ginning, neither Macarthy nor any other gins driven by steam have been introduced. It is a prevalent belief that those interested in the Government Dharwar gin factory, which supplies saw-gins fitted with saws imported from England, would use their powerful interest in the district to prevent native cotton merchants from sending their cotton to be ginned in the steam ginning factories that may be established. It is now highly expedient that the Dharwar Government gin factory should be abolished, leaving the field open for private enterprise. From what has been done in Guzerat, there need not be the slightest apprehension that the field will long remain unoccupied.

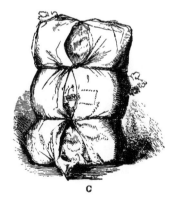

C

COTTON GINNING.

LENGTH OF STAPLE OF COTTON GROWN IN INDIA.

RESULTS OF DR. WATSON'S INVESTIGATIONS.

No.	Place of Growth.	Length of Staple. Inches and Decimals.			No.	Place of Growth.	Length of Staple. Inches and Decimals.		
		Min.	Max.	Mean.			Min.	Max.	Mean.
1	Surat	1·00	1·20	1·10	31	Dharwar	1·30	1·70	1·50
2	,, ...	·80	1·20	1·00	32	,, ...	1·00	1·20	1·10
3	,, ...	1·00	1·20	1·10	33	,, ...	·90	1·20	1·05
4	Guzerat	·90	1·20	1·05	34	Bunkapoor	·90	1·20	1·05
5	Broach	·80	1·00	·90	35	Dharwar	1·30	1·50	1·40
6	,, ...	·60	·90	·75	36	,, ...	·90	1·10	1·00
7	Dharwar	·90	1·20	1·05	37	,, ...	·80	1·00	·90
8	,, ...	·90	1·10	1·00	38	Coimbatore	1·10	1·20	1·15
9	,, ...	·90	1·10	1·00	39	,, ...	·80	1·10	·95
10	,, ...	·80	1·00	·90	40	Belgaum	·90	1·10	1·00
11	Tinnevelly	·90	1·20	1·05	41	,, ...	·80	1·10	·95
12	,, ...	·80	1·10	·95	42	Travancore	1·10	1·50	1·30
13	Trichinopoly ...	·60	1·00	·80	43	Mysore	·90	1·20	1·05
14	Tinnevelly	·60	·90	·75	44	Bolarum	·80	1·00	·90
15	Coimbatore	·70	1·00	·85	45	Sheopoor	·90	1·10	1·00
16	Candeish	·90	1·10	1·00	46	Tenasserim	1·10	1·30	1·20
17	Berar...............	·80	1·00	·90	47	Bolarum (Deccan)	·90	1·10	1·00
18	,, ...	·70	1·00	·85	48	Bengal, nr. Calcutta	1·00	1·30	1·15
19	Ahmednuggur...	·70	1·00	·85	49	Mysore	1·40	1·75	1·57
20	Belgaum	·70	·90	·80	50	Dharwar	1·50	1·70	1·60
21	Madras	·80	·90	·85	51	,, ...	1·50	1·80	1·65
22	Agra	·60	·80	·70	52	,, ...	1·50	1·70	1·60
23	Gwalior	·70	·90	·80	53	,, ...	1·50	1·80	1·65
24	Jeypoor	·70	·90	·80	54	,, ...	1·30	1·70	1·55
25	Jullunder Doab	·70	·80	·75	55	,, ...	1·40	1·70	1·55
26	Delhi...............	·50	·80	·65	56	,, (Hooblee)	1·40	1··0	1·60
27	Dharwar ',........	1·15	1·50	1·33	57	,, ...	1·40	1·60	1·50
28	Lingasoor	·90	1·20	1·05	58	,, ...	1·40	1·60	1·50
29	Guzerat	·90	1·80	1·10	59	,, ...	1·20	1·50	1·35
30	Dharwar	1·10	1·50	1·30	60	,, ...	·90	1·10	1·00

SUMMARY OF RESULTS.

Place of Growth.	Description of Cotton.	Length of Staple. Inches and Decimals.		
		Min.	Max.	Mean.
India................... {	Indigenous or Native......	·77	1·02	·89
	Exotic or American	·95	1·21	1·08
	Sea Island and Egyptian	1·36	1·65	1·50
United States	New Orleans, or Uplands	·88	1·16	1·02
Sea Island	Long Stapled	1·41	1·80	1·61
South America	Brazilian	1·03	1·31	1·17
Egypt	Egyptian	1·30	1·52	1·41

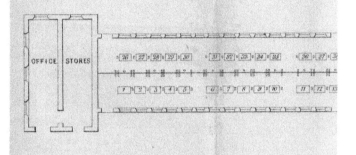

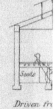

Seeds

Driven fr

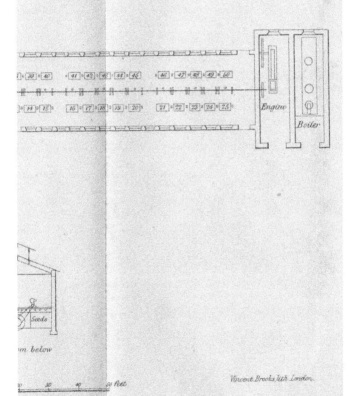

A.

GINNING FACTORY FOR 50 GINS.

Engine

Boiler

Seeds

m below

50 feet

Vincent Brooks, lith London.

ESTIMATE FOR A COTTON GINNING FACTORY.

£

25 Improved double-action Macarthy gins, 40 inches wide, with
self-feeding apparatus, and other recent improvements, complete 500
(Without improvements will cost much less.)
1 Horizontal steam-engine, high pressure, cylinder 14″, stroke
30″, with governor, fly-wheel, &c., complete ; 1 Cornish
boiler, 18 feet long, 5 feet diameter, complete, with mountings 300
Shafting for driving the gins, including wheels, pedestals, columns,
&c. 100
Leather belts (one large, from steam-engine) 50
Extras and miscellaneous articles, packing, and free delivery on
board 100

£1,050

DIMENSIONS OF A COTTON GINNING FACTORY
For 50 Gins.

	Length. Feet.	Breadth. Feet.	Square Feet.	Height.	Cubic Feet.
Ginning-room . .	160	25	4,000	15	60,000
Engine and Boiler-house	38	25	950	10	9,500
Total in square feet and cubic feet			4,950		69,500

REMARKS ON COTTON GINNING FACTORY.

GINNING is so important that cotton cleaned in a proper machine fetches a higher price than when cleaned in one of less perfect action.

The estimate for a cotton ginning factory is for 25 40-inch Macarthy gins, with all recent improvements. If the number of gins be increased, so much the better for economical working. No ginning factory ought to have a less number than 20 gins. The cost of the 25 gins, with self-feeding apparatus, and other improvements, with steam-engine, boiler, shafting, packing, and free delivery, comes to £1,050. From this estimate the cost of any number of gins, with the necessary amount of motive power for driving them, may be calculated. In some ginning factories, to open the tufts of cotton opening machines are used. Two will be required for 25 gins, and will not cost more than £50.

A steam-engine, two horse-power indicated, will turn three gins 40 inches wide. The space occupied by each gin will be 5 feet by 2 feet 10 inches ; packed in two cases they will weigh 7¾ cwt., and measure 30 cubic feet.

The yield of *clean* cotton from seed cotton per each gin will be from 20 to 25 lbs. per hour ; the better the quality of cotton, the greater will be the product. If the yield is 25 lbs. for Surat cotton, for Dharwar New Orleans it will be 35 lbs. per hour. In Guzerat one native churka, worked by two men, turns out 12 lbs. clean cotton and 24 lbs. seeds in a day.

Gins made at the Dharwar government factory, fitted with saws imported from England, cost about 173 rupees with 18 saws; for a gin with 10 saws, 100 rupees, which requires eight men to work it, and produces 200 lbs. of clean fibre per day.

It is not denied that at times the saws cut up the staple so badly that, notwithstanding the superior quality of Dharwar cotton, no higher price is realised than for the ordinary indigenous cotton. The Macarthy gin is adapted for cleaning cotton either short or long stapled, while the saw-gin is applicable for long stapled cotton only.

There is no doubt that by the use of the Macarthy gin the seeds are separated without in the least injuring the fibre, whereas with a saw-gin, even in long stapled Dharwar cotton, the condition of the fibre is not so perfect. Samples of cotton cleaned by the Macarthy and saw-gins respectively, have been submitted to the Manchester Cotton Supply Association; that cleaned by the former has been valued higher than the same quality of cotton cleaned by the latter.

By the application to Macarthy gins of a patent feeding apparatus, saving has been effected in labour; as one person can attend to more machines, while at the same time the production is also increased. The objection to expensive leather rollers has been obviated by the patent fibre roller, which it is asserted possesses greater durability, and an uniform regularity of surface, though in some factories it is not at all approved.

In America the Macarthy gin is used for the Sea Island cotton. It also cleans New Orleans cotton with less injury than the saw-gin; but for general purposes the Macarthy is not used in America, as it does a less quantity than the saw-gin used in the States.

Two patent gins have been lately constructed; and it has been asserted that by the peculiar action the fibre of cotton is protected from the tearing action of the saw-gin, and that the full length of the staple is taken off neatly; but these patent gins have not been yet tried on a large scale in a ginning factory.

Plan A of the cotton ginning factory shows the arrangement of fifty gins, the way the gins are driven by power derived from the steam-engine, and the room where the seeds fall from the gins during the process of cotton ginning. The engine and boiler are placed at the end of the main building. The gins may be driven from below, as shown on plan, or from top. In the case of a small factory of 25 gins, the plan of driving the gins direct will be preferable.

At the important cotton centre of Broach, in Guzerat, one of the ginning companies established is the Victoria Company, which works 80 double-action Macarthy gins, driven with engines of 25 horse-power. On an average, 62,000 lbs. of seed cotton or kuppas are cleared here every day. The Moffussil Company works 74 gins of the same description, with two engines of 15 horse-power each. The Anglo-Indian Company's factory contains 70 gins, Macarthy, with two engines of 25 horse-power each. There is another with 80 gins, driven by a 30 horse-power steam-engine; besides, other new gin factories have been projected, as they are paying so well. At Verumgaam, in the establishment of the Guzerat Company, there are 80 Macarthy gins worked by two engines of 20 horse-power. Each gin cleans 200 lbs. cotton daily. At Omrawatee the Moffussil Press Company works 80 Macarthy gins, and turns out 16,000 lbs. of cleaned cotton in ten hours, as stated in official reports.

The profits to be realised from ginning cotton by machinery with recent improvements may be judged from this fact, that two enterprising Europeans built a cotton ginning factory at Broach, commenced working during the season of 1865, and realised a profit of more than 35 per cent. on the capital laid out. The field for introducing improved gins worked by steam-power is so extensive, that it may be safely calculated that the profits on the average, under good management, will not be less than 15 per cent.

QUANTITY OF COTTON BALES IMPORTED INTO GREAT BRITAIN.

From	1866.	1865.	1864.	1863.	1862.	1861.
Bombay	1,206,660	956,886	1,043,378	898,900	915,170	906,640
Madras	294,372	177,882	173,648	177,000	124,200	79,950
Bengal	346,727	131,757	182,488	153,000	30,070	10
China and Japan	18,844	141,610	399,074	161,800	2,980	......
America	1,162,745	461,927	197,776	131,900	71,750	1,841,600
Brazil..............	407,646	340,261	212,192	137,900	133,810	100,000
Egypt..............	167,451	333,575	257,102	204,270	131,750	96,840
Turkey, &c......	32,770	80,303	61,793	44,430	14,800	960
West India, &c.	111,826	131,120	59,645	23,000	20,470	9,700
Total.........	3,749,041	2,755,321	2,587,096	1,932,200	1,445,000	3,035,700

AVERAGE WEIGHT OF COTTON BALES.

Descriptions.	1866.	1865.	1864.	1863.	1862.
Bombay	383	390	390	390	390
Madras	300	300	300	300	300
Bengal	298	300	300	300	300
China and Japan	326	240	240	200	200
America	441	423	438	440	435
Brazil	174	160	180	180	180
Egypt	490	492	500	470	450
Turkey, &c.......	340	350	355	344	...
W. India Islands	180	180	200	200	200
Average	362	355	347	364	369

WEEKLY AVERAGE CONSUMPTION IN EUROPE,
IN BALES.*

	1866.	1865.	1864.	1863.	1862.	1860.
Great Britain...	46,885	39,115	30,855	25,067	22,035	50,633
France............	11,808	10,943	7,808	6,577	5,981	11,942
Holland	3,538	2,039	2,115	2,654	1,519	2,250
Belgium	1,385	1,442	423	711	346	1,231
Germany.........	7,750	4,923	3,442	3,096	1,961	5,904
Trieste............	981	1,077	538	500	673	1,482
Genoa	288	519	327	461	250	1,385
Spain	2,442	1,788	1,750	2,096	1,577	2,039
Russia	5,654	4,662	4,288	3,538	2,384	6,211
Total.........	80,731	66,308	51,576	44,900	36,726	83,077

* From Board of Trade Returns, and Messrs. Ellison and Heywood's Circular.

WEIGHT AND VALUE OF COTTON IMPORTED INTO
GREAT BRITAIN.*

Year.	Weight in cwts.	Computed value. £.	Average price per lb.
1866	12,295,803	77,521,406	13·51d.
1865	8,731.949	66,032,000	16 2d.
1864	7,975,935	78,203,000	20·12d.
1863	5,948,422	56,277,000	20·27d.
1862	4,678,333	31,093,045	14·24d.
1861	11,225,178	38,653,398	7·37d.
1860	12,419,016	35,756,889	6·16d.
1859	10,946,331	34,559,636	6·76d.
1858	9,235,198	30,106,968	6·98d.
1857	8,654,633	29,288,827	7·25d.

* From Board of Trade Returns, and Messrs. Ellison and Heywood's Circular.

COTTON MANUFACTURES.

INTRODUCTION.

THE rapid growth and present magnitude of the cotton
manufacture are unprecedented phenomena in the history of industry. This manufacture now forms one
of the principal trades carried on in Great Britain,
affording an advantageous field for the accumulation and
employment of millions upon millions of capital. It
has contributed in no common degree to raise the
British nation to the high and conspicuous place she
now occupies. Nor is it too much to say that it was
the wealth and energy derived from the cotton manufacture that have given England strength to sustain
burdens which would have crushed any other nation.

Little more than a hundred years ago, cotton as an
article of commerce was scarcely known in Great

Britain. The import of the raw material into Liverpool was limited to a few bales. The first cotton mill that was built in England was so small and primitive that the machinery was turned by two asses walking round an axis; it gave employment to ten girls. The application of an admirable and simple contrivance by Richard Arkwright, of rollers for disengaging and laying parallel the cotton fibres, formed the basis of spinning machinery. Arkwright, Hargreaves, and Crompton were virtually the founders of the cotton manufacture, seeing that they determined with wonderful skill and ingenuity the principle of almost every machine now found in cotton mills, by successive improvements in which the production of yarn has increased three hundred fold. Arkwright's first mill was built in 1771, and was driven by water-power; but in 1790 James Watt's steam-engine was substituted, and, by its regular uniform motion, greatly improved the processes and imparted new life to the cotton manufacture. What had been a domestic occupation was now carried from cottages to factories, and conducted in one series of operations upon better mechanical principles and at less expense. Larger establishments were subsequently formed; the workmen were obliged to be more regular in their attendance, earned more wages, and lived better.

Improvements were also introduced in weaving. On the weaving loom, in 1800, the most industrious workman could produce in a week of sixty hours no more than four pieces of cotton cloth; whereas with the improved loom twenty-six pieces of the same fabric were produced, and that at a much less cost in working. But the important advantages of sizing and weaving by steam-power were not at once recognised. In

1813 there were no more than 2,400 looms worked by steam, but in seven years the number increased to 14,000, in 1830 to 60,000, in 1833 to 100,000; while the number of looms now working in Great Britain is more than 400,000. By the invention of the "loose reed," the loom for weaving cloth was worked with one hundred per cent. more speed than before; by the taking-up motion the cloth was rolled after being made; and by the use of change-wheels any number of picks in one inch of cloth could be put in with the greatest nicety and by self-acting mechanism. From 1830 to 1860 the productive power of cotton machinery made wonderful progress; this progress is kept up and will be continued in Great Britain, by the keen competition to make the machinery as self-acting as possible, and to increase the productive power as well as the quality. The improved machinery has given such an impetus to the trade, and now such immense quantities of yarn and goods are produced, that the whole population of the world would scarcely be able to spin and weave it by the use of the old hand-spindle.

According to the latest official returns, in the year 1861 there were more than 2,470 cotton mills in Great Britain; since then factories have been built with astonishing rapidity. T. Bazley, Esq., M.P. for Manchester, in an article in the *Exchange*, calculated approximately that the spinning spindles existing in the trade in the year 1863 were 32,000,000; and the gross capital invested by Great Britain in the manufacture, directly and indirectly, at *eighty million pounds sterling*.

The persons engaged in the manufacture are enabled to contribute one-fourth part of the national revenue of Great Britain, or exceeding twelve million pounds sterling per annum in taxes to the State.

In 1860 Britain exported to India alone 241,978,364 pounds of yarn and cotton goods; her total exports were 2,776,218,427 yards of cotton cloth, besides 197,343,655 pounds of cotton twist and yarn. The total declared value of cotton exports for that year amounted to £52,012,380, after supplying the household requirements of its own population. The importance of the cotton manufacture may be estimated from the fact that in 1860 out of the total exports from Great Britain, 40 per cent. consisted of cotton goods exclusively, 20 per cent. of three other textile fabrics, woollen, linen, and silk, while the remaining 40 per cent. included every other article of commerce. Every year the export of cotton manufactures increases remarkably; and notwithstanding that in the year 1866 during a great portion of the year there was a severe monetary crisis, its value, according to the Board of Trade returns, increased to £60,800,000.

Not only in Great Britain, but in almost all other countries in Europe, cotton mills have been built and fitted with the most improved spinning and weaving machinery. Gigantic establishments have been erected for making machinery for spinning and weaving cotton. One single firm near Manchester employs more than five thousand hands, to whom £250,000 are paid each year for wages alone. It is calculated that in it could be produced every week entire fittings and furnishings for a mill of 20,000 spindles for spinning and looms for weaving cloth.

But Great Britain has long ceased to possess exclusive knowledge and skill in the art of manufacturing cotton by machinery. The progress of this manufacture in Europe will be seen by the following table :—

Year 1860.	No. of Spindles.
Great Britain	30,387,267
France	4,000,000
Germany	2,000,000
Russia	2,000,000
Austria	1,500,000
Switzerland	1,300,000
Italy	500,000
Belgium	500,000

In 1866, the number of spindles working in Great Britain was 36,000,000, according to Mr. Ashworth, who read a paper on the subject before the Social Science Congress; he added that every day 10,000,000 yards come out from English looms.

India has reason to be ashamed that with its vast territories, with 200 millions of population all wearing almost entirely cotton fabrics, with raw material near at hand, with every facility in common with other countries for getting machinery with the latest improvements from England, with labour comparatively cheaper than in Europe—that in such a country, and with such facilities, no more spindles are running at the present moment than are now in the poorest third-rate countries in Europe — countries where no cotton is produced, where machinery for working is not made, where skilled labour is also wanting, and where also, as in Russia, the climate is unfavourable, the heat being excessive at one time, while at another the thermometer stands very much below zero.

The idea of a spinning and weaving factory in India is due, not to a native, but to a Frenchman, M. Des-bassyns, who was Administrator General in the French settlement of Pondicherry in the years 1826-28. At first the French Government granted large subsidies, and this branch of industry has prospered in Pondicherry. In Calcutta one or two cotton factories were erected, but through bad management did not make any progress. In Bombay, the first cotton mill, owned by "The Bombay Spinning and Weaving Company," did not commence working till February, 1856; and though it has paid handsome dividends, the number of cotton mills in Bombay, which place may be considered the head-quarters of cotton manufacturing by improved machinery in India, may almost be counted on the fingers' ends. There are not more than twelve cotton factories—a less number even than that established in Lancashire by working men on co-operative principles.

There is no doubt that that which impeded the progress of cotton manufactures in India was the rudeness and tediousness of the modes of working. In several places it is still carried on by the natives with the rudest and cheapest apparatus, just as their forefathers did before them. For thousands of years no improvement whatever was made in the art of fabricating cotton into cloth; whereas in England the improvements made during the last age have so economised labour as to enable one man to do the work of a hundred. It has been well said that England is more indebted to Arkwright and Watt for her triumphs than to Nelson and Wellington.

The cotton manufacture in India possesses elements of the highest commercial advantage — because the product is in demand over all parts of India and China, the raw material is near at hand, the processes of manufacture are improving every day, and it could be conducted under very advantageous conditions. It indicates one direction in which industrial activity and commercial prosperity may be successfully pushed.

THE TEXTILE MANUFACTURES AND THE COSTUMES OF THE PEOPLE OF INDIA.

(Extracts from Notices of a Work by J. Forbes Watson, M.A., M.D., F.R.A.S., &c., Reporter on the Products of India to the Secretary of State for India in Council. Printed for the India Office, 1866.)

From the *London Times*, January, 1867.

On no point, perhaps, in regard to India, has greater ignorance prevailed in England than in respect to the clothing of her vast population. We have heard and read of rich silken robes, and of gorgeous cloths of gold and silver tissue; and many of us may have seen and felt portions of that delicate filmy Dacca muslin which formed the envied dress of the belle of sixty years ago, or the final cravat which was tied so elaborately over the pad which encircled the neck of our great grandfather. When the East India Company traded to India, they im-ported some species of manufactures consigned from their "factories," and sold them in limited quantities. They were rarities and luxuries, and, generally,

very expensive. England then received " bandana "
handkerchiefs, calicoes, and some chintzes. Among
some of the treasured hoards of ancient ladies may be
remembered quilts called " pallempores," which were
spread over beds in summer time; as, also, among
gentlemen, were those real Indian " bandanas." But
these did not represent Indian manufactures, and were
goods made up especially for the English market by
Indian weavers. The gentlemen of India did not use
bandanas, and even the calicoes sent to England were of
a stouter and thicker texture than those used by the
people of the country, and were also woven to suit
English requirements. As to gold and silver cloths
and tissues, brocades, the richer descriptions of silk,
and the muslin fabrics in ordinary wear, they were not
needed in England; few beyond those who had seen
them in India knew of them at all, or cared to know
how they were woven or how used, or attempted to in-
vestigate the native arts of spinning, weaving, and dyeing—the manipulation, in fact, of products which could
not be imitated. . . . . Specimens of each fabric, particulars of which are detailed and printed with each,
are bound up in 18 volumes of folio size, and of themselves form valuable museums of practical reference.
We have had an opportunity of examining these
volumes, and found them, not only complete in every
respect, but forming a collection of most beautiful and
interesting samples of native manufactures. . . . . There
are 700 of these samples, all cut to the size of the folio
page, showing not only the texture, but the patterns of
centres of fabrics, with their ends and borders of silk,
cotton, or gold thread. The number of threads in the
warp and woof have been counted and are detailed, and
the length and breadth of each article, scarf, waistcloth,
piece, or whatever it may be, whether to be worn as a
garment, or part of a costume, or cut into a tunic, a
bodice, or any other portion of male or female apparel.
. . . . The range extends from the coarsest and lowest

priced cotton materials up to the richest and most elaborate specimens of gold and silver brocades, silks, satins, and muslins, plain, or figured with silver or gold thread.

From the *Manchester Examiner*, September, 1866.

By direction of the Secretary of State in Council for India a very large, and no less valuable, collection of samples of native textile manufactures has just been forwarded as a present to the Manchester Chamber of Commerce. Some idea of the copiousness of this collection can be formed when it is stated that it fills about eighteen volumes of two feet in length by one and a-half feet in breadth, and a thickness in proportion. These samples, which number upwards of 1,000, have been prepared at the India Museum, under the supervision of Dr. Forbes Watson, who has been engaged for some time past in reporting for the Government on the products of India. The specimens are in cotton, silk, satin, and woollen, and the patterns are so classified as to represent nearly all the uses, from turbans to gauze pantaloons, which such textiles may be required to serve. The first volume is entirely occupied with turban patterns. These are of all qualities, from the common bleached cotton from the handloom of Bhurtpore to the finest texture from Cashmere. Samples in use among the higher classes in Sinde are given in great variety. One of the richest of these measures $23\frac{1}{2}$ yards in length, by 13 inches in width, and weighs only 1 lb. 2 oz. . . . . Turbans in dyed cotton are equally well represented. There is one of remarkable richness, with two gold stripes down the centre of the principal end (extending about four yards). The weight of this with the necessary padding would be 3 lbs. It was manufactured at Oodipoor, in Rajpootana. In other samples of equal richness the colours are laid on with a stamp, and though padding has to be used in making them up, the weight, all told,

is less than 7 oz. The costliest specimen of all is from
Madras. In this sample the material, which is a mixture of silk and cotton, is fringed with gold thread.
Others have deep gold borders, and are valued in some
cases as high as £4. The assortments for men's gar-
ments are very numerous. In most of these samples
the two pieces are woven as usual in one length, with
a " fag " between, to permit of their being readily separated. The longer portion is worn round the body, and
the shorter one over the shoulders. The length of the
longer is from 4 to 6 yards, that of the shorter from 2½
to 4 yards, the average width being about 1½ yard, and
the price according to material. Cotton of fine light
texture, and of the above dimensions, is set down at
£1 10s. the garment; and, if common texture, the price
is from 3s. upwards. Samples of cotton worn by the
Lubbays, an industrious class of fishermen and Maho-
metan "merchants" on the Madras coast, are shown
from 2s. 7½d. to about 10s.; unbleached ditto, from
1s. 3d.; boys' ditto, length 1 yd. 3 in., width 19 in.,
and weight 2 oz., appears to be so low that it is not
priced at all. Vol. III. is chiefly composed of scarf patterns. Some of these are exceedingly bright in colour,
being woven in half-widths, with a border on one side;
two of these are afterwards sewn together so as to form
a complete scarf, with a border on each side. The
highest-priced sample is one which costs nearly £8.
Some are in silk, others in silk and cotton, others
(cotton) with silk borders and ends, and they show an
endless variety in colours and quality. There are also
samples of the " dhotee," a kind of scarf worn round the
loins. The volume ends with samples of cotton kess, a
coarse material used by the natives as a covering for
themselves as well as for horses. The length of one of
these samples is 3 yards 12 inches, width 46 in., weight
2 lbs. 2½ oz., and price 3s. 6d. Fabrics for domestic use
include samples of cotton palempore, or bed-cover, from
Bengal; these vary in price from 2s. 6d. to 6s., and

show great ingenuity of workmanship. Among the piece
goods are samples of cotton print known as chintz. . . . .
There are samples of the tent, railway, or sleeping rug.
These are from Upper Assam, and weigh from 6 lbs. to
7 lbs. The Sepoy "regulation rug," of which also there
are many specimens, appears as if it would wear for a
century. Some of these are from Madras and others
from Bengal. The "regulation" price is 2s. The pieces
for women's clothing show a dazzling variety of colour.
Crimson, puce, green, and pink are blended in true
barbaric splendour. Some of the finest muslins for
scarfs weigh only 6 oz. for 10 yards in length, and 1 ft.
4 in. in width. The price of these is from £1 upwards.
The chief place of their manufacture is Madras. The
silk tartans from Tanjore form also a conspicuous collection. The kincobs, mushroos, and silk brocades appear
to carry decorative manufacture almost to its height.
The kincobs, which are of satin, and used for ladies'
petticoats, are adorned with gold flowers. These wares
are principally from Trichinopoly, and cost from £3 to
£5. The brocades, with flowers of white silk, are from
the Deccan, though bought in Madras. The mushroos
have a silk surface, with a cotton back, and also bear
loom-embroidered flowers. These latter samples cost
about £2 for 5 yards in length by 30 in. in width, the
weight being about 1½ lb. Most costly samples are
massively flowered in gold, with silk stripes ; and others,
yet costlier, from Hyderabad, show a marvellously striking arrangement of colours, in wavy stripes of rich yellow
and gold with pink and white. The silk for trouserings
is of the thinnest and lightest texture. Nine yards of
it weigh scarcely so many ounces. Textiles of similar
fineness are shown in gauze.

COTTON SPINNING FACTORY.

MANUFACTURING PROCESSES.

(1.) *Cotton Opener.*—At one end of this machine cotton is spread upon an apron made of a series of thin narrow wooden bars fixed at their ends to two strips of leather. The cotton is conveyed to a pair of fluted rollers, and then to a series of revolving cylinders

furnished with rows of teeth, making more than a thousand revolutions per minute; the teeth seize the cotton and draw the locks apart. The dust falls within the casing, and is carried by a self-acting fan to the dust-room outside the mill. The cotton, after being opened and cleaned, falls out at the other end of the machine.

(2.) *Scutching and Lapping Machine.*—In this there is an arrangement for feeding a regular supply of cotton, which passes to the feed rollers, then to a set of toothed cylinders, then to a second and third set, revolving at a higher rate than the first, by which the fleece of cotton becomes reduced to one-third the thickness as first supplied to the machine. The next operation in the

same machine is to form the fleece into a roll or lap. By passing it to four calender rollers, the fleece receives three compressions, and is formed into a kind of felt, which winds itself upon an iron rod; this rod is removed when full, and is replaced by another.

(3.) *Carding Engine.*—The cotton laps are placed at the back of this machine, where the fibres are combed and straightened, and certain very slight impurities removed which otherwise would give roughness to the yarn. The carding engine combines self-stripping dirt-rollers, a main cylinder, with other rollers and clearers, the surface of which is covered with fine wire brush cloth. As these rollers and cylinders rotate in opposite directions, the cotton is seized by both sets of teeth, one

pulling one way, and the other the other. The fleece of downy cotton, which is now transparent, is detached by a blade, then contracted within a funnel, passed through rollers, and received in the form of a riband in a revolving can.

(4.) *Drawing Frame.*—In this frame six or more ends

of cotton slivers or ribands from the card-cans are passed through four pairs of rollers, revolving at different speeds, the speed of the front and that of the back pairs being as one to six. The result is that the slivers are drawn out to six times the original length, forming a single web, which is passed to and deposited in a revolv-

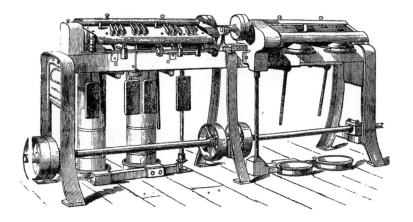

ing tin can. The centres of the rollers in this frame are adjusted to suit the length of staple of the cotton fibres. The cotton as it leaves the drawing frame is in the form of a loose, porous cord, the fibres of which are parallel.

(5) *Slubbing Frame.*—By passing through the rollers in this frame, the loose porous cord is further increased in length fivefold. In front of the machine are spindles in a double row. The spindle is a round steel rod with two arms or flyers, which fit to the top. Upon these spindles hollow tubes of wood, called bobbins, are placed. The two revolve at different rates. The cotton, as delivered from the rollers, is partially spun or twisted by the revolution of the spindles; it passes through the

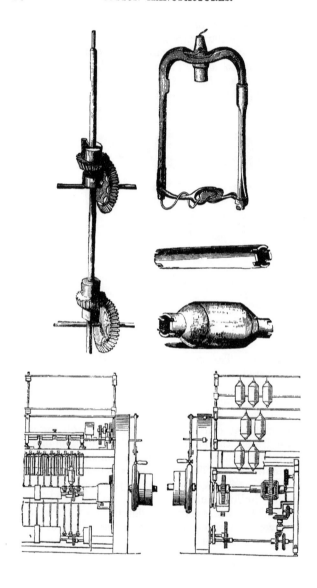

hollow legs of the flyers, and is wound upon the bobbins or wooden tubes.

(6.) *Intermediate Frame.*—The bobbins from the slubbing frame are taken to the back of the intermediate frame.   Two ends from two bobbins are then doubled by passing through another series of rollers, and joining, drawing, twisting, and winding them upon smaller bobbins.   The intermediate frame operates in the same

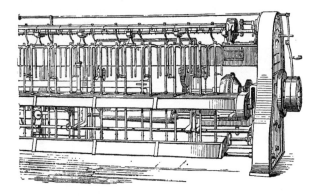

manner as the slubbing frame.   In mills where the yarn is not intended to be made of a superior quality, it is sometimes dispensed with ; but in spinning throstle yarn, the intermediate frame is essential.

(7.) *Roving Frame.*—The bobbins from the last frame are brought to this machine, where the cotton is passed through rollers, twisted, and finally wound upon still smaller bobbins.   To ensure a regular winding upon the bobbin, a presser is attached to the lower end of one of the flyer legs, and it is through a hole in the outer end of this presser that the roving passes in being wound upon the bobbin.   The duty of the hands who attend these frames consists in joining any sliver that may break, removing the full bobbins, replacing

empty ones, and stopping the machine for these purposes.

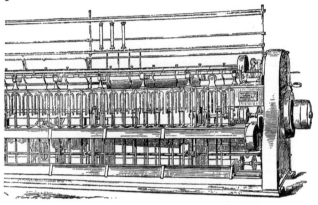

(8.) *Throstle Frame.*—The roving bobbins in this
machine are placed on the top, and the roving passed

through three pairs of rollers. The rail has an up-and-
down motion ; the small bobbin is moved up and down

the spindle, past the eyelet of the flyer, and the yarn is distributed equally upon it on winding. The yarn made on this machine is called throstle yarn, or water twist, which, being twisted hard, is used generally for warp. The throstle is used for "30's" and downwards —sometimes for "40's," but not for a higher or finer number.

(9.) *Self-acting Mule.*—This machine operates at the same stage as the throstle, but producing a different

kind of yarn. It consists of two principal parts. One is stationary, containing the rovings, the rollers for the elongation of the yarn, and wheel-work for governing the general movements. The other part consists of a carriage upon an iron rail, which holds the spindles, and traverses about five feet backwards and forwards. The yarn uncoils itself from the spindle down in a conical form called a "cop." Every movement in this machine is automatic. The mule is applicable for spinning principally weft, and also what is called "mock water," and any number could be spun.

(10.) *Reels.*—When yarn is *not* required for weaving, in the factory, but intended for sale, then it is reeled on these machines either from the throstle bobbins or mule cops, and wound off into measured lengths of 840 yards each, called *hanks*. The reel itself is six-sided, a yard and a half in circumference, on which the yarn winds; when it has completed eighty turns a check is

struck, showing that a rap of 120 yards has been formed. Seven of these make a hank of 840 yards. The girl who minds this simple machine ties the hanks round, slips them off, and proceeds to wind another set.

(11.) *Bundling.*—Yarn is sold in hanks, each containing a length of 840 yards. However fine the yarn may be, the same length is made to the hank, so that the quality or fineness is indicated by the number of hanks which make a pound in weight. Number 20 *water twist* means a coarse yarn of twenty hanks to the pound. Throstle yarn, as just mentioned, is called water twist, and is used for *warp* or longitudinal threads of cloth ; whereas mule yarn is used for *weft* or cross threads. In the bundling press a bundle of ten pounds is formed. Thirty of these, pressed in an hydraulic press, form a bale of 300 pounds for the Bombay market ; for other parts of India and for China fifty bundles of ten pounds each are made into a bale.

COTTON WEAVING FACTORY.

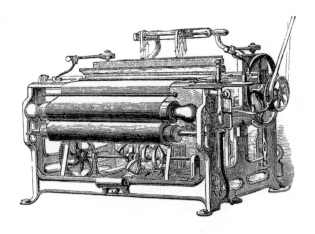

MANUFACTURING PROCESSES.

(1.) *Winding Machine.*—When the yarn is not intended for sale, but is made for weaving into cloth in
the same factory, then the warp yarn is brought to the
winding machine, where the threads are unwound, and
passed through minute slits, sufficiently wide to admit
the thread so long as it is of proper thickness, but in-
tercepting the thick parts. The threads are also passed
through brushes, which clear them of dirt. Then the
yarn is laid on bobbins, and there is an arrangement by
which the bobbins can be filled up in any form.

(2.) *Warping Machine.*—This is used for the purpose of winding the yarn or thread from bobbins on to beams, and to lay the threads alongside of each other in one parallel line. It has an arrangement by which undue tension is avoided, and breakages thereby obviated; but if a thread breaks, it is easily discovered and reunited. One of the oldest methods still practised

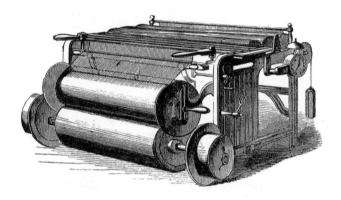

in India and China is to draw out the threads in the open fields. It is a very primitive process, no doubt, and is not at all countenanced in Europe now-a-days.

(3.) *Sizing Machine.*—As the threads in weaving are subject to considerable friction, it is necessary to give them a dressing of size or paste, which is done in this machine. The threads from six beams are all run together into boiling size, and passed between squeezing rollers, which press the size into the thread and strengthen it considerably. Then the yarn is dried by passing it over drying cylinders made of copper or sheet iron, and heated by steam-pipes. The yarn is finally wound uniformly on weaver's beams, and as the roll enlarges, the drag on the yarn is equalised.

E

(4.) *Looms.*—The beam of *warp* yarn is placed on the loom for weaving cloth, and each thread is passed through the teeth of an instrument called the *reed,* which is set in a movable swing-frame. One set of

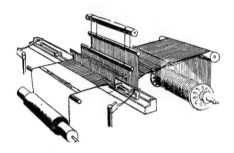

threads is raised and the other depressed, so as to make an opening or shed from one side of the warp to the other. In this opening a boat-shaped piece

of wood, called a *shuttle,* containing in its hollow the *weft* yarn, runs freely to and fro. The waft passes out from a hole as the shuttle moves along (more than 180

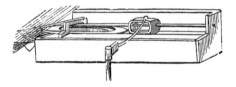

times a minute), and the weft thread is driven in. Each cast of the shuttle adds to the woven fabric by the breadth of a thread. As the weaving proceeds, the finished cloth winds itself upon a cloth beam.

(5.) *Folding Machines*, or Plaiting and Measuring Machines, driven by steam power, measure the cloth and lay it in folds after it comes from the loom, with the greatest regularity and precision, and effect a saving as compared with the old hooking system. The hydraulic press is used for pressing the cloth and making up into bales, the sizes of which vary according to the requirements of different countries.

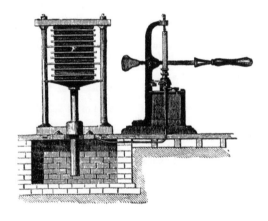

ESTIMATE FOR A COTTON SPINNING FACTORY OF
10,640 SPINDLES.

Machinery as shown on PLAN B.

            £

2 Opening machines, with 1 and 3 cylinders and fans. 1 scutcher and lap machine 250

14 Patent automatic carding engines, 40 inches on the wire, with cylinder stripping apparatus and patent doffer comb, 7 rollers, 5 clearers, coiler and can motion. (If the cards are ordinary, more than 20 will be required.) 980

3 Drawing-frames—1 of 3 heads and 4 deliveries, and 2 of 3 heads and 6 deliveries each ; 4 rows of rollers, coilers, and stop motions 370

3 Patent slubbing-frames, 46 spindles each, for bobbins 10 × 5 310

6 Intermediate frames—2 of 46, and 4 of 64 spindles each, bobbins 10 × 5 and 10 × 4½ 706

8 Roving-frames, 104 spindles each, bobbins 7 × 3½ . . 998

12 Self-acting mules, 672 spindles each, with double-boss rollers, backshafts, driving apparatus, complete . . 1,780

10 Throstles, 200 spindles each, top and under clearers, complete 758

4 Throstle and 16 mule reels, 3 bundling presses, and 1 hydraulic press. (Packing and free delivery included.) 790

1 Pair of horizontal compound steam-engines, high-pressure cylinder 18½ in., low-pressure cylinder 30 in. diameter, stroke 5 feet, with steam jackets, fitted with apparatus for cutting-off steam at different portions of the stroke, with air-pump, condenser, plates for bottom of engine-foundation, and with several improvements not to be found in ordinary engines. (Without improvements the cost will be about £300 less.) 1,500

3 Wrought-iron boilers, 25 feet long, 6 feet diameter, made of the best Staffordshire plates, complete, with steam-pipes and safety-valves, water and steam gauges, with duplicate articles 800

Green's economiser, 96 pipes, with scrapers and spare pipes 340

               Carried forward . . . 9,582

	£
Brought forward . . .	9,582
Floor-beams for engine-house, hand-rails, ornamental pillars for cranks, brass injectors, valves and pipes for connection with wells and reservoirs outside, &c. . . .	150
Mill-gearing, consisting of wheels bored, turned, grooved, and fitted, with 300 feet of shafting in several lengths as per plan, 115 self-lubricating pedestals, wall-boxes, brackets, fluted columns, axle drums, pulleys, heavy girders for blowing-room, cotton steam-hoist, iron staircase, iron sliding-doors, doors for entrances and warehouse, and all necessary duplicate articles, including packing and delivery	1,500
Tools for mechanics' shops, including a lathe, bed 13 feet, drilling-machine, set of drills, set of taps and dies, driving tackle, &c., including packing	175
Ironmongery, including wrap-reel, scales, roller-covering apparatus, roller-ending machine, assorted files, hammers, vices, chisels, pliers, bellows, emery-cloth, spirit-levels, fillet tacks, &c.	200
Strapping for driving all machinery, laces, roller-skins for covering rollers, roller and clearer cloth, &c. . .	450
Card-clothing, for main cylinder, doffer, take-in rollers, strippers, and clearers	148
Bobbins for slubbing, intermediate, roving, and throstles; creel-pegs, &c., painted and varnished, and made out of solid timber well seasoned, including 25 per cent. extra	330
Tinwork—including 2 oil cisterns of 200 gallons, with taps, drainers, indicator; cans for shafts, cards, and other machinery	98
Firebricks for boilers, if required	100
Miscellaneous articles, including grinding-machine for cards, card and other brushes, banding for throstles, felt, &c. .	222
	£12,955

ESTIMATE FOR A COTTON WEAVING FACTORY OF 200 LOOMS.

Machinery as shown on PLAN C.

	£
2 Winding machines, 200 spindles each, arranged to wind either from mule cops or throstle bobbins, complete . . .	135
3 Warping machines, for 500 bobbins each, with indicators, creel pegs, stopping, reversing, and measuring motions, &c. .	60
Sizing machine, with copper rollers, creel with 6 beams, copper drying-cylinders, size box, condensing boxes, expanding and contracting lathe, &c.	170
100 Looms to weave calicoes, shirtings, printing cloths, &c., 40 inches wide in the "reed space," with self-acting fluted roller temples, weft-stopping motion, positive taking-up motion, break, complete with all improvements, either fast or loose reed	950
100 Looms, 35 inches reed space, in every other respect similar to the above	900
1 Folding machine, 1 cloth press, 4 looming-frames, 2 drawing-in-frames, 1 patent heald knitting machine . . .	129
All extras and accessories, including yarn beams and flanges, boxwood shuttles, weft cans, dressers, flannel, bobbins, combs, sets of healds, &c., with packing and free delivery on board	828
	£3,172

NOTE.—If the looms are not driven by the engine shown on Plan B, a 15 horse-power engine will be required, which, with boiler, shafting, &c., will cost about £1,000.

SUMMARY OF ESTIMATE FOR COTTON SPINNING FACTORY.

	£
Cotton spinning machinery	6,942
Steam-engines, boilers, gearing, &c. . .	4,290
Tools for mechanics' shop	175
Ironmongery	200
Strapping	450
Card-clothing	148
Bobbins	330
Tin-work	98
Firebricks	100
Miscellaneous	222
	£12,955

(See " Remarks on Cotton Spinning and Weaving Factories.")

SUMMARY OF ESTIMATE FOR COTTON WEAVING FACTORY.

	£
Cop-winding machines	135
Warping machines	60
Sizing machine	170
Looms	1,850
Folding machine, &c.	129
Extras and accessories	540
Packing and delivery	288
	£3,172

(See " Remarks on Cotton Spinning and Weaving Factories.")

DIMENSIONS OF COTTON FACTORIES.

PLAN D,

or B and C plans, of spinning and weaving combined as one plan, containing 2,000 throstle and 8,064 mule spindles, and 200 looms.
Scale : 50 feet = 1 inch.

	Length, feet.	Breadth, feet.	Square feet.	Height, feet.	Cubic feet.
Front wing, for blowing-room, engine-house, 2 entrances, mechanics' shop, with upper story . . .	125	35	4,375	20	87,500
Side wing, for preparation machinery and throstles .	165	65	10,725	12	128,700
Side wing, for self-acting mules	165	65	10,725	12	128,700
Intermediate wing, for preparation machinery for weaving	125	35	4,375	10	43,750
Wing for 100 looms . . .	105	42	4,410	10	44,100
Wing for 100 looms . . .	105	42	4,410	10	44,100
Boilers and economiser . .	40	32	1,280	10	12,600
Warehouse	32	20	640	14	8,960
Total of square feet and cubic feet			40,940		498,410

PLAN E,

containing 4,752 throstle and 5,184 mule spindles, and 150 looms.
Scale : 40 feet = 1 inch.

Total length,　272 feet.
　,,　　breadth, 148　,,
　,,　　height,　12　,,
　,,　　square feet, 40,256.
　,,　　cubic feet, 483,072.

(See Note on Cotton Factory Plans, p. 65.)

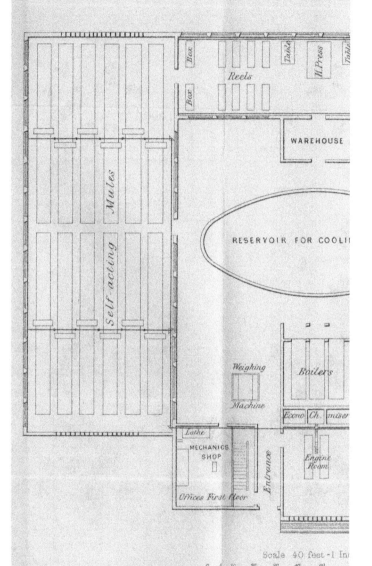

PLAN OF A COTTON SP

Box

Table

H. Press

Table

Box

Reels

WAREHOUSE

Mules

Self-acting

RESERVOIR FOR COOLI

Weighing

Boilers

Machine

Econo

Ch.

miser

Lathe

MECHANICS
SHOP

Entrance

Engine
Room

Offices First floor

Scale 40 feet-1 In

0   5   10   20   30   40   50

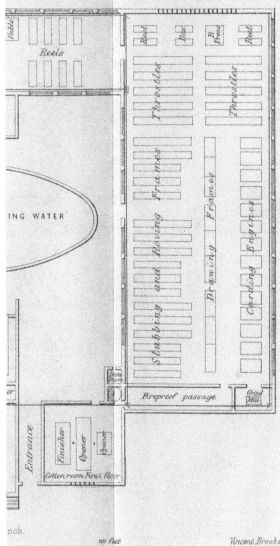

Reels

ING WATER

Reels

Throstles

Throstles

Stubbing and Roving Frames

Drawing Frames

Carding Engines

B Head

Dust Room

Fireproof passage

Grind Mill

Entrance

Finisher

Opener

Opener

Cotton room First floor

100 feet

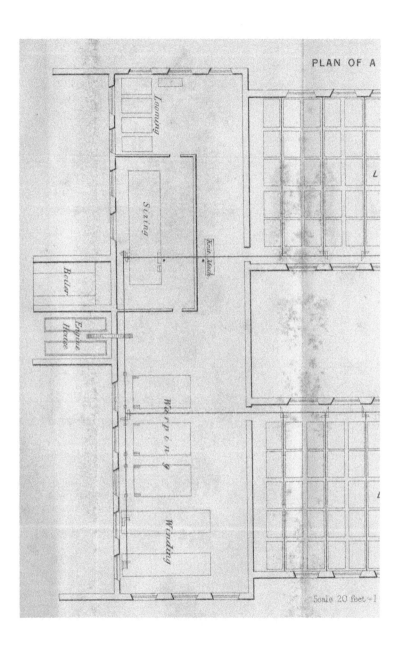

PLAN OF A

Laundry

Sizing

Iron. Mach.

Boiler

Engine House

Warping

Winding

L

L

Scale 20 feet = 1

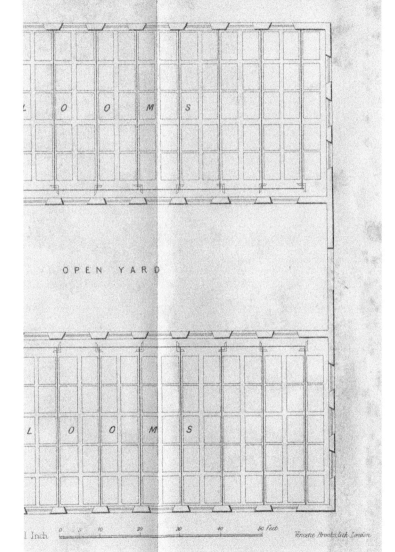

L   O   O   M   S

OPEN   YARD

L   O   O   M   S

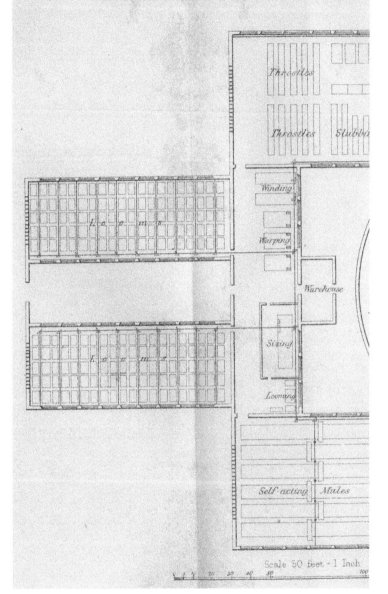

Throstles

Throstles   Slubbi

Winding

Warping

Warehouse

Sizing

Looming

L o o m s

L o o m s

Self-acting Mules

Scale 50 feet = 1 Inch

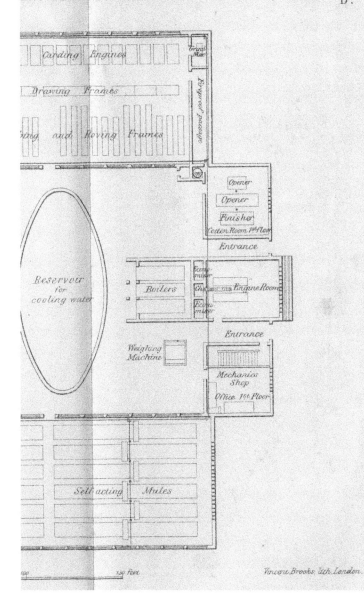

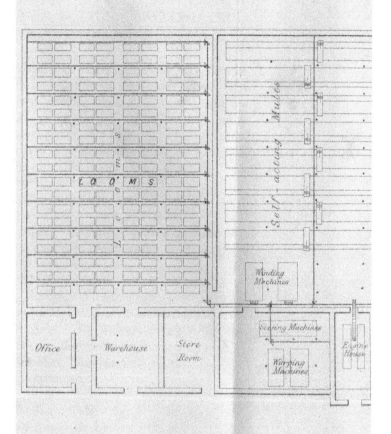

PLAN OF A COTTON SPINNING &

LOOMS

Self - acting Mules

Winding Machines

Sizing Machines

Warping Machines

Office

Warehouse

Store Room

Engine House

Scale 40 feet =

& WEAVING MILL, BOMBAY.

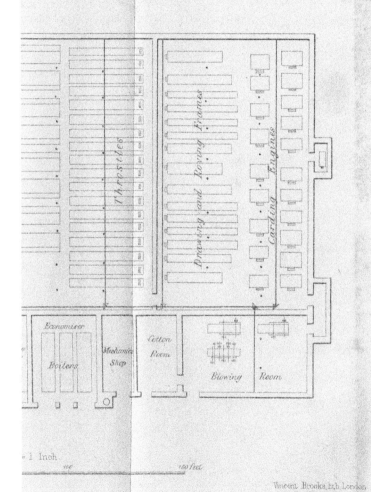

REMARKS ON COTTON SPINNING AND WEAVING FACTORIES.

THERE is no country in the world where the population exceeds in number that of India or China. The natives are almost entirely clothed in cotton, and the clothing consists generally of coarse fabrics. The quantity consumed per annum has been estimated from 5 to 20 pounds per head of population. Now, taking only India, with a population of 200 millions, and an average consumption per head of population no more than 3 pounds, the annual consumption would amount to 600 million pounds, quite irrespective of China and other parts of Asia. To supply this very extensive demand of cotton goods for the teeming millions offers a field of vast industrial enterprise, in a country where the raw material is produced, and where labour is cheaper than in Europe. No other branch of industry offers so much scope for investment of capital as cotton manufactures in Asia. By manufacturing with improved machinery on the factory system, rather than with rude implements, the comfort of the millions would be greatly promoted.

The principal use to which Indian cotton was put in cotton mills in England, before the American war, was as a mixture with American cotton. But now Indian cotton is mixed with every variety. In spinning fine numbers, for which the demand is very limited, there are some obstacles; but in spinning coarse numbers, and weaving them into cloth adapted for India and China, there is no difficulty. During the cotton famine in England the Manchester Cotton Supply Association obtained from numerous English spinners a large collection of yarns, ranging from No. 4's up even to No. 70's, all guaranteed exclusively from East Indian cotton. A quantity of cloth manufactured from Broach cotton was bleached, dyed, and printed by some of the principal houses in the trade, in styles ranging from a single colour to a twenty-colour machine furniture pattern, which proved the capability of Indian cotton for the production of goods of a quality for which the demand is so extensive not only in India and China, but in almost all parts of Asia.

In the Bombay Presidency the demand is chiefly for No. 20's mule and water twist; while in Madras, Calcutta, and China, it is generally 40's mule yarn. In some mills in Bombay No 40's is spun from Indian cotton regularly. For weaving calicoes No. 32's is a common sized yarn. For candle wicks and coarse counterpanes, such low numbers as two hanks to the pound are also manufactured.

Machinery for which the estimate is given is adapted for spinning yarn from No. 20's to 40's.

Plan B of the spinning factory shows arrangement for little more than 10,000 spindles. At Ahmedabad a mill was started with only 2,500 spindles; in Bombay in two cases with 5,000 spindles; but to work with full advantage, 10,000 spindles ought to be the least number in a factory—the more the better.

The steam power required for driving the machinery, according to the firm who made the engines for the

Arkwright Mill, is that a nominal horse power will drive 200 throstle spindles with preparation ; or 400 mule spindles and preparation ; or 20 looms ; the engines indicating, when fully loaded, four times their nominal power, spinning and weaving medium numbers. But it will be a wise provision in every factory to have more driving power than less.

Production from the spinning machinery for which the estimate is given, and shown on plans B and D, in a week of sixty hours :—

8,000 mule spindles, spinning No. 40's . . . . 5,000 lbs.
2,000 throstle spindles, spinning No. 20's . . . . 2,000 lbs.

The higher or finer the number of twist, the less the production will be. The self-acting mule can be made applicable both for weft or cross threads and warp, or longitudinal threads of cloth ; but the warp in mule not being so hard twisted as in throstle spinning, there is an advantage in throstle over mule warp in weaving, particularly T cloths and long-cloths, but not much in light goods. Throstle spinning requires nearly twice the amount of steam·power to drive it as mule spinning—while a mule requires much more space than a throstle, which will be seen at once on the plans ; so that there is hardly any advantage in throstle over mule spinning.

The cost of spinning yarn in Bombay is just the same as in a cotton mill in Lancashire, for No. 20's, which is the principal number in demand in the Bombay Presidency. The waste in working Indian cotton, when not properly cleaned, is, on the average, about 20 per cent., more or less, according to the class of cotton used ; in American cotton it is a little less.

Looms and other machinery for weaving yarn into

cloth, for which the estimate is given, and shown on plans C and D, are adapted to weave shirtings, prints, Madapollams, T cloths, and other cotton goods in most extensive demand in India and China. The counts of yarn most used for weaving this class of piece-goods are 24's warp and 32's weft.

The quality of cloth is ascertained with some precision generally by a small magnifying glass, by counting the number of threads in a given space.

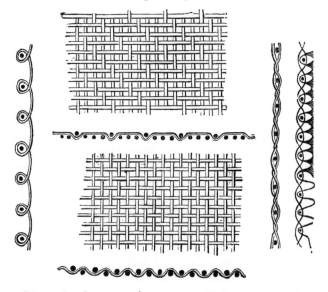

The only changes of pattern which can be readily produced by plain weaving are stripes or checks—the former generally depending upon the colours of the warp, and the latter upon the colours of both warp and weft. Thus, stripes in the direction of the cloth may be produced by using warp of various colours, or a warp composed of threads of different sizes and substances ; stripes across the web may be formed by using shuttles containing various colours and substances; checkered patterns, by varying both warp and weft ; and figures, to a certain extent, by raising and depressing alternately certain portions of the warp. *Twills* are formed by causing the thread of the weft to pass alternately over four and one of the threads of the warp, and performing the reverse in its return.

The Wages in England, in the cotton factories, for the men who mind the mule spinning, are from 20*s.* to 25*s.* ; the boys earn, according to their age, from 6*s.* to 14*s.* per week. In the weaving department, the winders earn about 25*s.* per week ; and the weaver, generally a woman, who attends two or sometimes three looms, earns from 12*s.* to 14*s.* per week. Except in a few cases, the wages to the hands are paid according to one standard list of prices, and in proportion to the amount of work they turn off. This system has been very lately introduced into Bombay mills, only in weaving, but ought to be introduced in the spinning department as well, as it is a great stimulus to the hands to work steadily. In the Bombay factories, the adult hands receive from 10 to 12 rupees; boys and girls, from 4 to 8 ; and the women, from 6 to 8 rupees per month. Labour in Bombay is dearer than in other parts of Asia, where in some places it is 50 per cent. lower.

The capital required for the spinning mill of 10,000 spindles as per approximate estimate of machinery amounts to £12,955, to which add for freight and insurance about 15 per cent., and for erecting machinery about 10 per cent. For mill-house with iron roof, galvanised sheets, glass windows, two rupees per square foot, more or less according to the price of building materials and wages at the place where it may be intended to build the factory, and of which an exact estimate may be made from the dimensions given of the plans of factories. For the spinning factory, including dead stock, &c., a capital of £30,000, and if it be intended to add 200 looms at the same time, a capital of £40,000, will suffice for all ordinary purposes.

Profits.—The first mill that commenced working in Bombay was that of "The Bombay Spinning and Weaving Company," with a capital of 500,000 rupees, divided into shares of 2,000 rupees each. The following is the official memorandum of profits derived by that Company, the manager of which, a Parsee gentleman, receives a clear commission of 5 per cent. on the proceeds of yarn sold ; a due reward in this case, as he was the first to introduce cotton spinning into Bombay with improved machinery ; and which good example has been followed, though slowly, but on the whole with success, by the establishment of a dozen companies for cotton spinning and weaving.

DIVIDENDS PAID BY THE BOMBAY SPINNING AND WEAVING COMPANY.

When paid.	Dividend.	Capital.	Amounts of Dividend.	Remarks.
		Rs.	Rs.	
Dec., 1858	First	500,000	600	Commenced work-
June, 1859	Second	,,	400	ing February, 1856.
Dec., 1859	Third	,,	600	
Dec., 1863	Fourth	,,	690	From 690 received
June, 1864	Fifth	550,000	700	back Rs.500 to in-
June, 1865	Sixth	,,	400	crease the capital.

DADAHOY PESTANJEE, Secretary.

Bombay, 17th April, 1866.

Another establishment, which was organised in the year 1861, and called "The Bombay United Spinning and Weaving Company," with a capital of 900,000 rupees, has declared dividends as follows :—

10 per cent., first dividend for the half year ending June, 1863.
20 per cent., second dividend for the half year ending December, 1863.
12 per cent., third dividend for the half year ending June, 1864.
12 per cent., fourth dividend for the half year ending December, 1864.
No dividend for the year 1865.
12 per cent. for the year ending 1866.

" The Bombay Alliance Spinning Company," during the six months ending 1866, made a profit of £16,491 on their paid-up capital of £125,000.

Regarding Bombay cotton mills, at a meeting of the Directors of the Manchester Chamber of Commerce, held in April, 1867, one of the directors, Mr. Cassels, read a letter from his firm, as follows :—

" Bombay, March, 1867.

"The longcloths, T cloths, and domestics produced by these concerns have been gaining in favour with consumers in all districts where they have been selling during the last two or three years, and are now preferred to Lancashire makes of the same class, mainly because the former, although a rougher, is an honest article, and wears better than the latter. Experience has taught the natives that, owing to the extent the majority of your manufacturers have been adulterating their cloth with size of late years, Lancashire cloth deteriorates very much in the first washing. Of the total quantity of these goods sold in the Bazaar, from 50 to 75 per cent. are of local make. All the Bombay mills are working full time; and, as they can sell their productions at paying rates faster than they can manufacture, they are making good profits.

(Signed)   "PEEL, CASSELS, & Co."

The losses which some of the Bombay cotton mills have suffered in 1865-66, were from taking large advances from banks at a very heavy interest, accumulating yarn and goods from month to month, and then selling them at a much lower price than in the first instance.

It must be borne in mind, that as in every other trade and profession, so in cotton mills and all other factories, the profits and losses will fluctuate more or less according to the price of raw materials and manufactured goods, to good management, and so on. But on the whole, investment of capital in factories will be as safe as any other, and less liable to fluctuations than bank shares, or shares of reclamation companies.

NOTE ON PLANS OF COTTON FACTORIES.

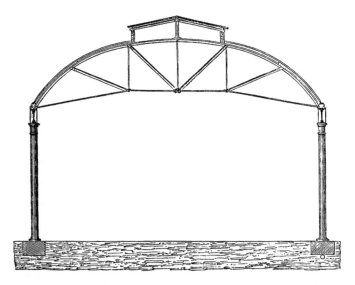

Four plans, marked B, C, D, and E, of factories for spinning and weaving cotton, have been given; and, though necessarily reduced to a small scale from large plans, will answer all the purposes, as showing the disposition of machinery, the way the machinery is driven, and also the space the buildings for factories will occupy.

B. *Plan of Cotton Spinning Mill* for 2,000 throstle and 8,064 mule spindles, and the other necessary machinery for working it; for which the estimate has been given.—All the machinery is placed on the ground floor. In this plan some important improvements have been adopted, suggested by actual experience, and which have been introduced in almost all the other plans of factories contained in this work. The principal feature is this: that instead of one building, the mill is divided into separate wings, but still connected with one another. The extreme breadth of the wings is 65 feet, but that is an exception—in several cases it is only 30 feet, and even less. The light in the mill is derived solely from *side* windows inserted in the side walls. In the mills erected in

Bombay this is not the case. There the light is entirely from *top* windows. The main advantage of the adoption of wings is, that the mill being lighted by side windows facing each other, and not from the roof, the expense of several thousand feet of glass windows absolutely required for the mills as erected in Bombay, as shown in plan E, will be saved to a great extent; and the light from side windows will be quite ample for the purpose of working the machinery. The ventilation in the mill by having ventilators, and that opposite each other, will be far better than is the case at present; and the nuisance of gutters will be entirely obviated by having the iron roof made of one span, without a central gutter.

C. *Plan of Cotton Weaving Factory*—contains 200 looms and all the preparation machinery for weaving, for which estimates have been given.—The looms are placed in two wings, 100 in each wing, and the breadth is so regulated that the light from side windows would be quite ample. The plan C may be considered as a continuation of plan B of spinning factory, if it is intended to use the yarn for weaving. The reeling machines which are required in a spinning factory not being wanted in weaving, the place in the spinning factory occupied by reels is here filled up with preparation machinery required for weaving.

D. *Plan of Spinning and Weaving Factory*—includes all the machinery for spinning as well as for weaving, and shows at one view the two plans B, C, combined as one whole. In the front wing of this factory, just in the middle, is placed the steam-engine house; the boilers for supplying steam, and the fuel economiser being immediately behind, and if desired may be entirely separated from the wing by leaving a free passage between the engine-house and boilers. There are two entrances on two sides of the engine-house; one would answer the purpose, but two would be better. At one end of the front wing is the blowing-room for cleaning cotton, and at the other end is the mechanics' shop, and staircase for going to the upper story over the front wing for raw cotton storeroom, offices, &c. If preferred, instead of the upper story, two rooms may be built at the two ends of the front wing, one for storing raw cotton, the other for offices. Two large side wings are attached to the front wing; one contains the carding, drawing, slubbing, and roving frames and throstles; the other the self-acting mules. The back wing corresponding to the front wing, contains preparation machinery for weaving, to which is connected two smaller wings filled with looms for weaving. The whole arrangement is such on plan D, that in spinning and weaving one process will follow the other with regularity, till the raw cotton will come in the shape of finished cloth into the warehouse.

Near the blowing-room is a fire-proof passage to the other wing, secured with iron doors, as required by the insurance companies. The cleaned cotton formed into laps will be taken through the passage to the large wing, there to be carded, drawn, twisted, and spun on throstle machines, or otherwise spun in the self-acting mules in the wing opposite, where the bobbins will be carried down by a special contrivance to save manual labour. Between the throstle and mule spinning-rooms there is an intermediate wing where the cotton yarn will be beamed, sized, and then passed to the two wings behind containing the looms for being woven. The finished cloth will be deposited in the adjoining warehouse.

At first sight it seems that this plan D occupies a great deal of space; but it does not, as the *open* yard in the centre is reserved for building one or two large reservoirs of water required for the purpose of condensing steam, and for storing coals, &c. This factory, with its projecting wings, with one story over the front wing, will also look far more imposing, possess even architectural beauty without any additional cost, and occupy less space for its size than any of the cotton mills built as yet in Bombay.

E. *Plan of Cotton Spinning and Weaving Mill* erected in the Bombay Presidency, illustrates the principle on which all the cotton mills have been built there. It contains 4,752 throstle spindles, 5,184 mule spindles, and 150 looms, with the other necessary preparation machinery for spinning and weaving. The building is in length 276 feet by 140 feet in breadth. Inside the building are iron columns shown on plan in .... dots, about 20 feet one way, and 10 feet the other way, forming what is called a bay. The breadth of the mill being 148 feet, the light admitted from the side windows is quite insufficient for the purpose of working the machinery. Hence for mills built on this principle it is absolutely necessary to derive the necessary light from a series of glass windows placed *on the top* of the roof, supported by girders resting on iron columns, the windows running right across the entire breadth of the mill on every alternate row of columns.

Many thousand square feet of glass windows are thus required even for a small mill. The iron columns being hollow, are used for taking away the rain water from the gutters, which run outside the mill by an underground gutter.

The height of all mills, where the machinery is on the ground floor, is generally 10 feet from the floor line: where the brickwork or masonry is cheap it may exceed this by a few feet; but that is quite arbitrary. The subject of factory buildings is treated of separately at page 290.

QUANTITY OF COTTON YARNS

EXPORTED FROM GREAT BRITAIN TO INDIA AND CHINA.*

To	1866.	1865.	1864.	1863.	1862.
	lbs.	lbs.	lbs.	lbs.	lbs.
Bombay	3,761,021	3,361,582	5,765,996	6,115,350	4,646,075
Madras	4,709,541	2,381,229	2,090,351	3,957,747	1,663,956
Calcutta	11,404,079	7,860,149	8,481,515	10,791,512	9,681,377
Singapore	1.759,954	1,297,300	1,172,718	1,991,936	1,593,371
Ceylon	782,376	335,400	176,546	612,226	448,440
China..............	4,741,350	1,078,208	1,961,813	2,251,154	3,214,059
Total, including other countries.	139,005,221	103,533,609	75,738,845	74,398,264	93,225,890

DECLARED VALUE OF COTTON YARNS

EXPORTED FROM GREAT BRITAIN TO INDIA AND CHINA.*

To	1866.	1865.	1864.	1863.	1862.
	£	£	£	£	£
Bombay	381,807	359,536	741,608	710,565	381,199
Madras	542,948	298,032	329,751	531,743	157,933
Calcutta	1,244,938	824,981	1,119,564	1,370,929	797,338
Singapore	173,046	127,181	.130,790	213,231	99,389
Ceylon	82,933	34,941	28,227	78,746	40,479
China..............	428,937	104,414	241,930	239,251	183,963
Total, including other countries.	13,700,404	10,342,737	9,096,209	8,063,128	6,202,240

* From the Board of Trade Returns.

QUANTITY OF COTTON PIECE GOODS
EXPORTED FROM GREAT BRITAIN TO INDIA AND CHINA.*

To	1866.	1865.	1864.	1863.	1862.
	Yards.	Yards.	Yards.	Yards.	Yards.
Bombay	155,341,732	163,810,271	196,165,111	245,570,066	170,843,749
Madras	22,043,668	18,934,880	14,190,874	18,438,591	10,058,058
Calcutta	368,455,276	321,035,019	228,993,986	244,072,772	282,600,808
Singapore........	59,422,703	43,217,688	24,980,447	29,060,076	32,726,886
Ceylon	26,429,513	15,594,561	12,645,684	22,222,525	18,432,697
China...............	188,610,971	126,160,887	73,462,152	46,454,793	80,636,465
Total, including other countries.	2,575,967,256	2,014,303,716	1,748,927,590	1,710,962,072	1,681,394,600

DECLARED VALUE OF COTTON GOODS
EXPORTED FROM GREAT BRITAIN TO INDIA AND CHINA.*

To	1866.	1865.	1864.	1863.	1862.
	£	£	£	£	£
Bombay	3,005,234	3,265,893	4,723,182	4,964,622	2,504,698
Madras	491,829	416,404	407,374	455,565	189,529
Calcutta	7,138,941	6,127,628	5,149,227	5,017,537	4,484,287
Singapore........	1,331,808	918,966	653,357	701,618	567,488
Ceylon	836,610	384,476	413,536	569,642	310,538
China...............	4,420,168	2,779,408	2,010,025	1,162,505	1,269,988
Total, including other countries.	57,829,440	44,876,363	43,887,387	37,633,535	28,562,466

* From the Board of Trade Returns.

EXPORT OF INDIAN MANUFACTURED GOODS TO FOREIGN COUNTRIES,
IN EACH OF THE YEARS ENDING 30TH APRIL.†

Year.	Value. £	Year.	Value. £
1865	1,043,960	1858	809,183
1864	1,167,577	1857	882,241
1863	785,437	1856	779,647
1862	748,385	1855	817,103
1861	786,557	1854	769,345
1860	763,586	1853	930,877
1859	813,604	1852	819,049

† From Statistical Abstract presented to Parliament in 1867.

WEIGHT AND VALUE OF COTTON MANUFACTURES IN GREAT BRITAIN,
WITH THE COST OF PRODUCTION, AND THE BALANCE REMAINING FOR INTEREST OF CAPITAL AND PROFITS.*

	1866.	1865.	1864.	1863.	1862.	1861.	1860.
	lb.	lb.	lb.	lb.	lb.	lb.	lb.
Cotton Consumed............	890,721,000	718,651,000	561,196,000	476,445,000	449,821,000	1,005,477,000	1,079,321,000
Less Waste in Spinning......	115,793,000	100,611,000	78,567,000	71,466,000	76,469,000	105,575,000	113,328,000
Yarn Produced......	774,928,000	618,040,000	482,629,000	404,979,000	373,352,000	899,902,000	965,993,000
Exported in Yarn............	134,889,000	98,563,000	71,951,000	70,678,000	88,554,000	177,848,000	197,343,000
Do. in Piece Goods,&c.	490,713,000	377,357,000	332,048,000	321,561,000	324,128,000	406,284,000	542,770,000
Home Consumption, Stock	149,326,000	142,120,000	78,630,000	12,740,000	39,330,000	225,770,000	225,880,000
Total as above	774,928,000	618,040,000	482,629,000	404,979,000	373,352,000	899,902,000	965,993,000
	£	£	£	£	£	£	£
Value of Yarn Exported......	13,598,000	10,351,000	9,467,000	8,679,000	7,523,000	9,292,000	9,870,000
Do. Piece Goods, &c.	66,145,000	51,005,000	53,100,000	19,046,000	38,616,000	41,514,000	46,248,000
Home Consumption, &c. ...	23,020,000	21,910,000	13,710,000	2,070,000	3,413,000*	23,525,000	24,470,000
Total Value of Goods ..	102,763,000	83,266,000	76,307,090	59,795,000	42,726,000	74,331,000	80,588,000
Cost of Cotton Consumed ...	51,958,000	47,257,000	52,462,000	40,689,000	26,734,000	32,205,000	28,910,000
Wages and other Expenses.	31,288,000	23,850,000	18,680,000	15,690,000	14,520,000	31,360,000	33,600,000
Total Expenditure..	83,246,000	71,107,000	71,142,000	56,379,000	41,254,000	63,565,000	62,510,000
Balance left for Profits	19,517,000	12,159,000	5,165,000	3,416,000	1,472,000	10,766,000	18,078,000

* From Messrs. Ellison and Heywood's Circular.

BLEACHING AND FINISHING WORKS.

INTRODUCTION.

BLEACHING is the process by which the impurities and colouring matters in grey goods, cotton, silk, or other textile fabrics, are removed, and the cloth rendered nearly or quite white. In India it has been practised from ancient times, as well as in Egypt and other countries, where lime, ashes of plants, or potash and other substances, are used for bleaching. At one time woven goods made in England were sent to Holland to be bleached; where, after some preparatory processes of boiling and washing, the goods were spread out on the grassy fields, and exposed to the action of air, light, and moisture, for several weeks. In India the process was, and is up to the present time, very laborious and tedious. In Europe several attempts have been made to find out substances by which bleaching could be done well, and in a short time.

In 1785 it was discovered that a gas contained in common salt, called *chlorine*, possesses the property of destroying vegetable colours. This was a triumph; as by the application of this gas in bleaching, the work of several hands is reduced to a few; what required weeks is now performed in a single day, or a few hours, and consequently done at a mere fraction of the former expense. But chlorine gas, being poisonous, had an injurious effect upon the workpeople. A convenient method was soon, however, discovered of applying the gas in combination with lime, forming what is known by the name of *bleaching powder*, or chloride of lime. A solution of this is made in water; the gas, being chemically combined with lime, remains in combination, and does not escape into the air: while, on the other hand, the combination of the gas with lime does not prevent it from operating upon goods. It is now extensively employed for bleaching, and in conjunction with improved machinery for boiling and drying the goods, the whole process of bleaching is performed with the greatest ease and speed; thus entirely doing away with the necessity for large plots of ground or fields, and for the great amount of labour required by the old and primitive process.

There are several circumstances which interfere with

the whiteness of cotton and other goods. In spinning and weaving, the goods acquire certain impurities from the perspiration and dirt from the hands of the work-people. The fibres of cotton contain also a colouring matter naturally adhering to it, which gives to it the peculiar grey colour, but which seems to have no influence in giving strength to the fibres ; for the cloth is found as strong after the discharge of the colouring matter as before. The fibres are also covered with a resinous substance which prevents them from readily absorbing water. There is also the size or starch used in the threads before weaving. These and other impurities in the grey cloth are removed by the several processes in bleaching, now practised with success in most European countries.

In Great Britain there are hundreds of *calenderers*, whose business consists in finishing the goods after bleaching, and making the surface of the cloth compact, level, uniform, and glossy. This is cheaply and very expeditiously done by passing the cloth between heavy rollers, in machines worked by steam. In India, down to the present time, a glossy appearance is given to the cloth by rubbing over the surface with a shell by hand, as hard as the force of the wrist can bear. It is a most costly and laborious process, and not fit to be practised even in a half-civilised country. Asia is still following the old process. The wisdom and philosophy of its people consist in doing exactly what their wise forefathers did before them centuries ago ! But it is neither wisdom nor philosophy ; it is downright folly.

BLEACHING.

PRACTICAL PROCESSES.

(1.) *Singeing.*—The object is to remove the fibrous down or nap from the surface of the cloth, otherwise it will injure its appearance when finished. To facilitate subsequent operations, the pieces of cloth are first fastened together at the ends by a stitching machine, then wound upon a roll, and passed over a red-hot plate of copper in the singeing machine, at such a velocity that it singes or burns off the loose fibres but without injuring the cloth. Some goods are singed on one, and some on both sides.

(2.) *Boiling in Keir.*—To remove the grease the goods are boiled with lime, or other alkaline substances, in a keir or boiler, heated by means of steam pipes. It has contrivances to regulate the quantity of liquor; to spread it over the cloth; to ensure that the cloth

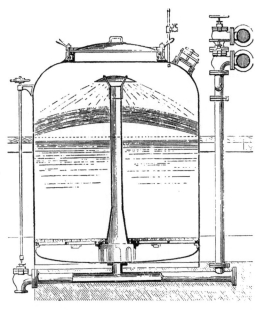

does not rise so high over the solution as to endanger its working; and, if such were to happen, to check the steam at once. The quantity of lime employed is various, but the usual allowance is three per cent. of the weight of cloth. The action of the lime makes the goods look darker in colour than before.

(3.) *Washing.*—The machine for this process has a trough for water, and two wooden squeezing rolls. The cloth passes round the roller in water, and is then squeezed by the rolls, which make 100 revolutions per minute. In the machine there are pegs to guide and to regulate the tension of the cloth, weights and levers for giving the desired pressure, water taps, &c. The

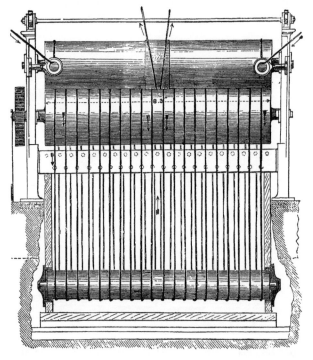

machine is capable of washing about 800 pieces of cloth, and requires 400 gallons of water per minute. Formerly dash-wheels were used for washing, but are now almost entirely given up.

(4.) *Chemic and Scouring Machine.*—This is similar in most respects to the washing machine, the only difference being that the rolls are smaller. It is called *chemic* on account of some acid being used in solution with water for washing. The acid sets free chlorine gas from the chloride of lime, and removes the colour from the cloth. If it is not sufficiently white, the operations of washing are repeated more or less till the

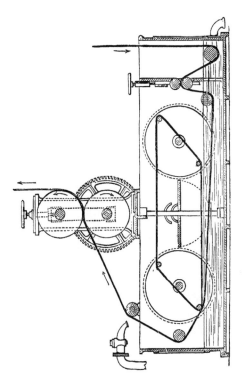

proper whiteness is obtained. After scouring or washing with acid, the cloth must be again washed as before.

To give an idea of the vast capabilities of some washing machines, the following fact, mentioned by an eminent firm, has been cited: that their machine "will wash 12,000 yards per hour for all bleaching purposes, and 6,000 yards for all kinds of dyeing purposes, and only equiring the attention of a person of twelve or fourteen years of age."

(5.) *Squeezing Machine.*—This is used for discharging the greater part of the liquid or water from the goods used in the process of bleaching. The *white* squeezer is the machine used for squeezing before the goods are dried. In some processes strong solutions

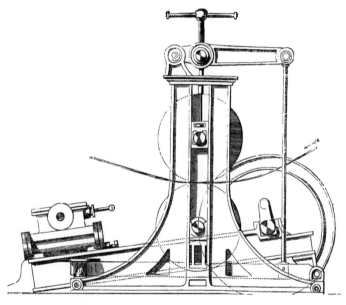

of chemicals are used, but by improved machinery the cloth is passed from one process to the other so rapidly that it has no injurious effect on the cloth. According to the class of goods to be bleached, the process of bleaching varies in details.

(6.) *Drying.*—The goods are dried on this machine. It has more than ten cylinders heated with steam. It is furnished with vacuum valves, an apparatus for folding the cloth, &c. The cloth, when passed over from one cylinder to another, becomes quite dry, but not smooth. If the cloth is bleached for the purposes of printing, no further operation is required; but if intended to be sold as white goods it is starched and glazed, to improve it in appearance and finish.

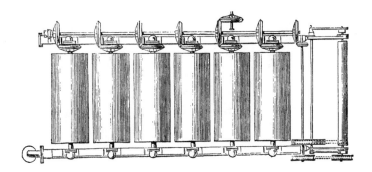

FINISHING.

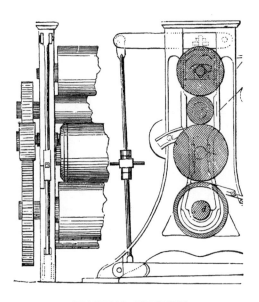

PRACTICAL PROCESSES.

(1.) *Mangling or Pressing.* — The bleached goods requiring to be finished are passed through mangles, consisting of brass and cotton rolls alternately. The wrinkled cloth is dragged through a cistern of water, and passed between rollers forced together by levers to which considerable weight is attached, where it is mangled and pressed, and the surface flattened and equalised. It is next wound upon a roller to be starched, which is the next process. Starching is sometimes carried to great excess to increase the bulk, and deceive the purchaser as to the quality of the cloth.

(2.) *Starching and Drying.*—The starching machine contains a trough for holding starch. A roller which dips into the starch lays it regularly and evenly on the cloth, while other rollers press out the superfluous portion. Flour paste is the substance usually employed,

and sometimes fine porcelain clay is mixed with it. In some cases, the starch is mixed with blue, to heighten the effect. After this the cloth is dried by the drying machine, consisting of hollow copper cylinders, heated by steam, and is then stretched upon the stretching machine.

(3.) *Glazing or Calendering.*—Many kinds of goods require glazing, for which purpose the cloth is first damped in a machine where a revolving brush throws water in a fine spray over it, after which the cloth is passed through a calender, consisting of rollers, on which a very powerful pressure is brought to bear; so that

G

the cloth becomes smooth and somewhat shining.   The
rollers for calendering or glazing are of iron, wood,

paper, &c., and require very nice workmanship in preparation.

(4.) *Making up.*—The cloth after being glazed by
the calenders is folded into lengths and forms according

to the different kinds of goods, or the particular market
for which they are designed.

DYEING.

INTRODUCTION.

THE art of dyeing has been practised from the most ancient times in the East Indies, Persia, China, Egypt, and Syria. In India, the mode of dyeing *Turkey red,* which is the most durable vegetable tint known, was first discovered. It was afterwards practised in other parts of Asia and Greece. In 1765 the French Government, convinced of its importance, published an account of the process. In England it was introduced in the eighteenth century, when a Turkey red dye-work was first established in Manchester. About the same time a Frenchman introduced it into Glasgow, at which place the greater part of the Turkey red dyeing in Great Britain is still carried on.

When indigo, the most useful and substantial of all dyes, so largely produced in Bengal, was first introduced into Europe, it met with great opposition. It was considered a dangerous drug, and called *food for the devil;* and by an Act of Parliament, passed in the reign of Queen Elizabeth, was ordered to be destroyed in every dye-house wherever it could be found. The use of logwood as a dye was also prohibited till the reign of Charles the Second. But the absurd prejudices against indigo and logwood soon gave way. By the discovery of the properties of the different dye-drugs, and especially by the knowledge of the modes in which the drugs combine with other substances, the art of dyeing has made very rapid progress; so that the European dyer has it now in his power to produce a variety of tints of great depth, durability, and lustre. The recent application of chemistry has also shown that the dyestuff formerly thrown away is available again to a great extent, by simply treating it with acids. But in India and other parts of Asia the art of dyeing is at a stand-still. The best means of extracting colouring principles, and the most effectual manner of applying them to cloth, as done in Europe, have not been yet introduced into a country where nature is so bountiful.

The greater number of colouring matters employed in dyeing are derived from plants. They are soluble in water,—much more so in hot than in cold water,—and readily attach themselves to the cloth to be dyed; but in washing with water the colour easily separates.

To unite the colour with the cloth another substance, termed *mordant*, is applied, which possesses the property of forming an insoluble compound with the colouring matter, thereby *fixing* the colour, which remains even after repeated washings in water. A mordant thus acts as a bond of union between the cloth to be dyed and the colouring matter. Common alum is an example of a mordant, and is extensively employed in dye-works for that purpose. The dye which is the product of combination between the mordant and the colour is not always the natural colour of the drug, seeing that a great variety of colours are capable of being produced by varying the kind of mordant. For example : if a piece of cloth be impregnated with alumina, and then passed through madder solution, it acquires rose tint. In the same dye, with a solution of iron as mordant, the cloth becomes dark brown or even black. With another mordant, the same infusion of madder yields a variety of tints between the most delicate pink and dark red.

There are instances also where the dye drug in its natural colour is fixed fast within the fibres of the cloth. Amongst this class of dyes is indigo. It is not soluble in water, but by mixing it with lime, and depriving it of part of its oxygen, it loses its colour, and becomes soluble in water. The cloth on being immersed in this solution, and afterwards exposed to air, imbibes oxygen, the original colour is restored, and it becomes indigo-blue, forming a permanent dye on the cloth.

Dyeing Cotton Yarns.—In dyeing cotton yarns, the yarn is brought to the dye-house in bundles of ten pounds, and after banding, it is boiled in water until thoroughly wet, and each roll put upon a wooden roller.

If the colour to be dyed be dark—such as brown, black,
purple, or deep blue, the yarn is now ready for the
dyeing operation; but if for light shades—such as pink,
&c., the yarn is bleached by boiling in bleaching solution previous to being dyed.   For blue no heat is
required in the dyeing vats.   Six vats constitute a set.

During the operations of dyeing there are certain
circumstances which have to be attended to in order to
facilitate the processes.   Thus with many colouring

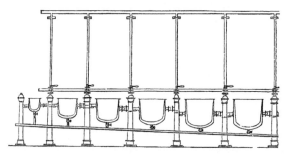

substances heat is necessary, not only for making the
solution, but also for its combination with the stuff or
mordant.   Decoctions of wood are always made in hot
liquor.   Colours dyed by munjeet or madder must be
done at a boiling heat during the whole process, or no
dye is effected.

The permanence of colours is another important
matter.   The term "fast colour" is generally given to
colours that will resist all ordinary washings.   Some
fast colours are immediately affected by acids or alkalies,
but Turkey red is not.

In *Turkey red dyeing* great excellence has been
attained in Europe, shades being produced of great brilliancy and fastness.   The bleached yarn is first soaked

in oil, then dipped in carbonate of soda, and exposed to the action of air and steam in a hot room. The yarn is passed next through a solution of nut-galls and a red mordant, successively, and is ready for dyeing, to effect which it is boiled for two or three hours in a vessel with madder root or munjeet. Lastly, the brilliancy of the colour is completed by boiling in soap solution. Fictitious Turkey reds are abundant in the market, and though approaching the genuine colour very closely, are unable to resist wear and washing.

The wonderful progress which Turkey red dyeing has made with European appliances, and with the assistance of chemistry, as compared with the mode in which it was carried on for centuries in Asia, and still extensively practised, renders it impossible to deny that material advantages would result by introducing improved machinery for that purpose. The table of shipments of coloured yarns and cloth from Great Britain will show clearly how extensive is the field for adopting improved methods and improved machinery in Asia.

DYE DRUGS PRODUCED IN ASIA.

Some dye-stuffs are yielded by nature in a state fit for immediate use, such as madder, logwood, sandalwood; but indigo, litmus, &c., undergo a process of preparation before being used for dyeing purposes.

It was observed that some of the munjeet or madder grown near Avignon, in France, where it was introduced by a Mahomedan, from Persia (whose bronze statue, erected to perpetuate his memory, may be seen on the top of the hill from the Avignon railway station), was found inferior in its richness and brilliancy of colour. A chemical examination of the soil was made, and the result showed that the soil was deficient in lime in comparison with some of the best madder farms. A dressing of lime was given to the soil, and it was found that the crop of madder yielded was inferior to none. The value of the dye-stuff depends a good deal on the care bestowed on its cultivation. Owing to the judicious manner in which the Chinese safflower is cultivated and collected, it contains far more of the fine red colouring matter; it is consequently worth four or five times as much as the best Bengal safflower. A short account of the principal dye-stuffs will not be out of place here.

Catechu is the dried juice of the khair tree of Bombay and Bengal. Its application in some styles of dyeing and printing has been of the greatest service. It has allowed a scope of design and variety of colouring by using different mordants, which has done much to extend the use of printed goods. It is largely used to give various shades of brown and other lighter colours springing from it. Bengal catechu is in flattish round lumps of a light brown colour outside, but Bombay catechu, or *Terra japonica,* is in square masses. Its colour is reddish brown, and it breaks unevenly.

Cochineal, employed for dyeing purposes, is a female insect feeding on nopal trees, where it remains fixed upon the sap till killed in hot water. There is only one male to about two hundred females. The season for rearing lasts seven months, during which the insects are collected three times by being brushed from the plant. They are then spread out, dried, and packed for the market. Cochineal is one of the most expensive dye drugs. It is easily soluble, and easily extracted by boiling the insects. The colours which are derived are red, pink, scarlet, and crimson.

Gall Nut is a valuable dyeing material. It is produced originally by the puncture of a little insect on the branches of the tree to deposit its egg. The juice of the tree collects round the egg, and hardening, forms the gall nut, which contains from 50 to 70 per cent. of tannic acid, valuable in dyeing black colour. Sumach and logwood are used for the same purpose.

Indigo is an important dyeing material, and is derived from plants. India produces nearly four-fifths of all the indigo consumed in the civilised world. The best quality from Bengal fetches from three to four times the price of the lowest quality. The method of dyeing in this colour is very simple ; and all the styles of work produced are cheap and low class. It possesses extraordinary stability, and therefore its chief consumption is among the poor classes. It seldom enters into high class work. The finest qualities of indigo in Great Britain are obtained from Bengal. Thousands of maunds are exported from Bengal every year, made in Jessore and Kishnaghaur. That made in Tyroot, Oude, and Benares, is inferior in quality to Bengal. There are four qualities of Bengal indigo, which by passing over into one another produce a number of intermediate colours. The Madras indigo is more light and friable than Bengal. When cotton goods are to be dyed of a uniform blue, neither bleaching nor washing is at all necessary. The size in the unbleached goods rather facilitates the dyeing process, in which a peculiar roller apparatus is employed.

Lac-dye yields the same colours as cochineal, and is prepared in India from a resinous substance which exudes from certain plants by the puncture of the insect, and which is called stick-lac. The tinctorial principle of lac is made by evaporating the infusion to dryness, and forming the residue into cakes. It is imported into Great Britain in two forms, called lac-lake and lac-dye, which contain about 50 per cent. of colouring matter.

Logwood.—The colouring principle is red, but it also forms a variety of shades between light purple and

black. The long pieces are rasped or cut by machinery
in the dye-house into small fragments fit for extracting
the colour. It is extensively used in black dye, as its
cost is less in England than that of galls, which give
the best and firmest colours.

Safflower is the flower of a plant growing in India
and other warm climates. It contains two colouring
matters, one worthless for dyeing, the other useful for
a fine red colour. It is largely employed for dyeing
cotton and silk. Safflower possesses the remarkable
and exceptional property of fixing itself readily and
firmly on cotton, without the intervention of a mordant.

Myrobalsam is the fruit of a tree which grows in
India, and is used in that country in dyeing, as a substitute for galls and sumach, and also for the same
purpose in Europe. It is exported from Bombay in
large quantities.

Madder, or munjeet, is sometimes known as *Bombay
roots*, and grows in many parts of the world. It is
most extensively used in dyeing and printing of cotton
goods for the production of a permanent bright red
colour. A form of madder containing more colouring
matter than the natural root is prepared, and called by
the name of *garancine*. The chief colours obtained
from madder, besides Turkey red, are madder purple,
madder pink, purplish black, and various shades of
brown.

ESTIMATES FOR BLEACH AND FINISHING WORKS.

Machinery for Bleach-house, as shown on PLAN F.

		£
1 Singe stove, copper singe plates, rollers, framing, complete	.	96
1 pair wrought-iron keirs, with cover, pipes, valves, complete. 1 washing machine, with 2 rolls, complete. 2 pairs of squeezers, with rolls, complete. 1 chemic machine, with trough rollers, complete. Top gearing and winches, for drawing from one machine to the other	.	374
1 Drying machine, with steam cylinders, pressure guage, safety-valves, taps, stands, troughs, complete	.	168
6 Steam-engines, with governor, valves, and gearing, complete, for driving all the above machinery; boiler, pipes, valves, and taps for bleach-house; and packing and free delivery on board	.	665

£1,303

Machinery for Finishing Works, as shown on PLAN F.

		£
2 Water mangles, 6 and 3 rolls, brass spreading-rollers, double arrangement of levers, warming cylinder, complete .	.	1,135
2 Starching mangles, brass rolls, &c., complete. 1 patent stretch-ing machine, with batching apparatus, &c. 1 stretching machine. 1 damping machine; 1 conroy, complete .	.	695
1 Finishing calender, with 5 rolls, with double batching ar-rangements, complete	.	455
7 Steam-engines, with governor, valves, and gearing, complete, for driving the above machines	.	600
1 Hydraulic press, ram 10″, with set of pumps, boiler, feed pump, blowing fan, heating cylinder, &c.; including packing and free delivery on board a vessel	.	750

£3,635

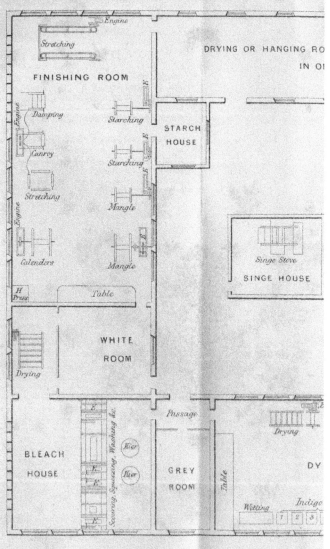

PLAN OF BLEACHING, DYEING A

DRYING OR HANGING RO

IN OI

Engine

Stretching

FINISHING ROOM

STARCH
HOUSE

Damping

Starching

Canroy

Starching

Stretching

Mangle

Singe Stove

SINGE HOUSE

Calenders

Mangle

H
Press

Table

WHITE
ROOM

Drying

Drying

Passage

BLEACH
HOUSE

Kier

Kier

GREY
ROOM

Table

DY

Scouring, Squeezing, Washing &c.

Wetting

Indigo

| 1 | 2 | 3 |

Scale 33 feet =

10 5 0   10   20   30   40   50

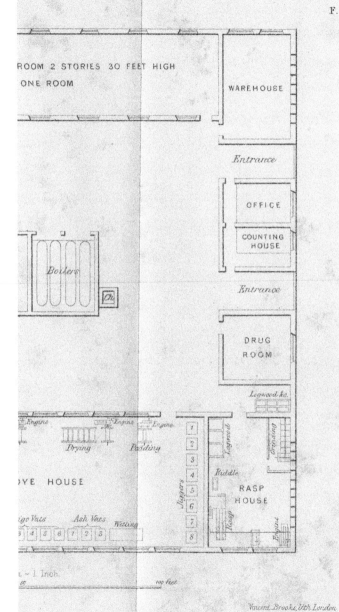

ROOM 2 STORIES 30 FEET HIGH
ONE ROOM

WAREHOUSE

Entrance

OFFICE

COUNTING HOUSE

Entrance

Boilers

Ch.

DRUG ROOM

Logwood &c.

Engine   *Engine*   *Engine*

Drying   *Padding*

Chamber   *Grinding*

Rialle

Lappers

| 1 |
| 2 |
| 3 |
| 4 |
| 5 |
| 6 |
| 7 |
| 8 |

RASP HOUSE

Rasp   *Engine*

YE HOUSE

Indigo Vats   *Ash Vats*   *Wetting*

| 3 | 4 | 5 | 6 | 1 | 2 | 3 |

t = 1 Inch.
50   100 feet

Vincent Brooks, Lith London

ESTIMATE FOR DYE HOUSE.

Machinery as shown on PLAN F.

£

2 Wetting cisterns, drawing rollers, fittings, &c. 8 iron jiggers, with reversing rollers, complete. 6 indigo vats. 3 ash vats, with rollers, including steam-engines for driving, complete   750
1 Padding machine, with 2 brass rolls, including steam-engine for driving, &c., complete   90
2 Drying machines, with steam cylinders. 2 mangles, including 2 steam-engines for driving, complete   532
Rasp for cutting dye-wood, with sets of knives, complete. 6 mills for grinding indigo ; grindstone, with frame, &c. 2 steam-engines for driving, complete. Boiler for supplying steam, complete ; including packing and free delivery on board a vessel   577

£1,949

DIMENSIONS OF BLEACHING, DYEING AND FINISHING WORKS.

Machinery as shown on PLAN F.

Scale, 33 feet = 1 inch.

	Length, feet.	Breadth, feet.	Square feet.	Height, feet.	Cubic feet.
Bleach house, with grey room . .	74	48	3,552	14	49,728
White room . .	31	53	1,643	14	23,002
Starch house . .	22	18	396	10	3,960
Finishing room .	100	53	5,300	14	74,200
Drying room . .	130	31	4,030	30	120,900
Dye house, &c. .	100	48	4,800	14	67,200
Rasp house . . .	48	37	1,776	14	24,864
Warehouse, &c. .	131	28	3,668	15	55,020
Singe and boiler house . .	60	28	1,680	10	16,800
Total of square feet and cubic feet			26,845		435,674

REMARKS ON BLEACH, FINISHING, AND DYE WORKS.

MACHINERY AS SHOWN ON PLAN F.

PLAN F, for Bleach, Finish, and Dye works, is so arranged that each of the three departments may be separated and made complete by itself if desired; or the three departments, so intimately connected with each other, may be combined as one whole, as shown on plan—forming complete works for bleaching, dyeing, and finishing goods at the same place. This would be preferable.

The bleach house is adapted to bleach more than *two thousand five hundred pieces of twenty yards each per week.* In bleaching, the length of the piece is of more importance than its weight, and it is by the length that the quantity that could be bleached is always calculated, and not by its weight.

The cost of bleaching in England, including profits of the bleacher, &c., is from eight to ten pence per piece of 38 yards, according to finish and width of cloth.

In the dye house, as shown on the same plan, one set of six indigo vats will dye fifteen pieces of fifty yards each *per day*, of the best genuine indigo blue; or double that quantity of medium quality, topped with Prussian blue ; or ninety pieces of common topped blue, topped with logwood. In facy dyeing, two jiggers will dye twenty pieces per hour of slate, drab, or lead colour ; or forty pieces of black, brown, &c. In fact, the quantity that can be dyed will depend on the quality and the style of colour.

Dyeing in indigo blue and other colours is extensively practised in India ; and it will be seen from the tables that large quantities of coloured yarn and cloth are shipped every year from Great Britain to India and China. There is ample room, therefore, for introducing improved machinery for dyeing purposes, as the stuffs used for dyeing are produced in large quantities in almost all parts of Asia.

The finishing room on Plan F, for starching and glazing the piece-goods, is so adapted as to finish the goods that may be either dyed or bleached in the bleach or dye house shown on same plan.

The cost of machinery for bleaching, dyeing, and finishing, including packing and free delivery on board, will not be more than £7,000, exclusive of freight, cost of buildings, &c. The total cost of the works complete, with reservoirs of water, will probably amount to about £17,500.

Bleaching, as well as dyeing, is intimately connected with *calico-printing,* so further remarks will be made under that head.

QUANTITY OF DYE-STUFFS IMPORTED INTO GREAT BRITAIN.*

	1866.	1865.	1864.	1863.	1862.
Indigo ....cwts.	74,256	66,506	76,214	85,395	69,589
Madder Roots .... ,,	345,052	237,352	314,926	355,681	299,873
Logwood ....tons	34,960	26,847	41,625	38,404	41,257
Cochineal................cwts.	30,721	24,260	23,396	26,120	22,760
Cutch ....tons	29,396	19,702	22,673	35,541	29,720
Terra Japonica .... ,,	2,434	2,120	2,980	2,020	2,069
Shumac .... ,,	12,845	13,588	12,292	12,807	13,819
Valonia .... ,,	11,862	10,015	11,143	19,094	11,125

VALUE OF INDIGO AND OTHER DYES EXPORTED FROM INDIA TO GREAT BRITAIN.†

Year.	Indigo.	Other Dyes.	Total Weight.	Total Value.
	£	£	lbs.	£
1865	1,860,141	80,354	1,946,495	1,940,495
1864	1,756,158	93,788	1,849,946	1,849,946
1863	2,126,870	80,287	24,721,452	2,207,105
1862	1,647,503	112,911	17,472,430	1,760,414
1861	1,886,525	203,042	25,271,411	2,089,567
1860	2,021,288	114,485	29,746,494	2,185,773
1859	2,118,016	121,279	26,709,126	2,239,295
1858	1,734,399	123,123	23,305,959	1,857,462
1857	1,937,907	87,151	14,062,284	2,025,058
1856	2,424,332	58,991	19,633,164	2,483,233
1855	1,701,825	115,427	14,866,473	1,817,254
1854	2,167,769	113,518	19,249,817	2,182,288
1853	1,809,685	100,559	16,151,261	1,909,119
1852	2,025,313	98,919	18,863,519	2,123,732
1851	1,980,896	102,431	17,387,933	2,083,335
1850	1,838,474	68,891	16,066,653	1,907,021

EXPORT OF COLOURED PIECE GOODS FROM GREAT BRITAIN TO INDIA AND CHINA.‡

Year.	Madras and Calcutta.	Bombay.	China.
	Yards.	Yards.	Yards.
1866	53,240,396	32,101,829	22,839,136
1865	55,892,289	32,763,232	9,254,667
1864	46,468,212	36,222,728	11,225,982
1863	40,476,360	43,692,979	12,139,413
1862	54,046,985	27,887,453	10,789,865
1861	62,083,760	38,916,737	17,981,519
1860	48,450,836	43,251,624	25,620,787
1859	66,153,829	37,887,090	14,755,082
1858	35,339,742	46,344,325	18,071,354
1857	38,824,224	24,175,156	14,308,353

* From Board of Trade Returns.   † From Statistical Abstracts.
‡ From Messrs. George Fraser, Son, & Co.'s Circular.

CALICO PRINTING.

INTRODUCTION.

CALICO PRINTING is the art of producing chintzes or other coloured patterns on cotton cloth, by printing in colours, or mordants, which become colours when sub-

H

sequently dyed.   India is the birthplace of this art,
which derives its name from Calicut, a seaport town on
the Malabar coast, where formerly the art was carried
on extensively; and large quantities of printed cottons
were exported to Europe from that place by the Dutch
and East India Companies.   But in England, at one
time, a class of people were so jealous of Indian chintzes
that an Act was passed by Parliament prohibiting the
introduction of the beautiful prints of India; this restriction "prevented the British ladies from attiring in
the becoming drapery of Hindostan."   This Act was
intended to protect the English wool and silk manufacturers from the competition of Indian goods.   But the
English became accustomed to the use of printed calicoes
and chintzes; and thus gradually a trade was established in England by printing on plain Indian calicoes,
which were up to that time admitted under a duty.   In
1712 the printing business which was first introduced
from India became rather extensive, so an excise duty
of threepence per square yard was levied; but this
duty retarded, and in many ways prevented, the progress of the art in Great Britain; so, after a short time,
it was quite repealed.

In India, for *bandana* handkerchiefs, the method
practised from time immemorial is to bind firmly with
thread all spots on the cloth to be protected from the
dye, and to remain white, while the rest of the cloth is
subjected to dyeing operations.   This most laborious
process was improved in Europe in 1811 by employing
a series of presses worked by steam, and bringing a
strong pressure on an engraved metal plate, which contains holes, through which lemon juice or citric acid is
made to pass, which dissolves the colour of the cloth

immediately. In this simple manner, by connecting a series of presses, 1600 pieces of 12 yards each are converted into *bandanas* in the short space of ten hours by the labour of four workmen only. Thus the Indian style has not only been copied, but far surpassed in cheapness, by the Europeans, with the aid of chemistry and machinery.

The art of printing calicoes has been improved steadily in Great Britain from time to time. The chief improvement was the invention of the cylinder or roller printing-machine, which has almost entirely superseded the hand block-printing process. Printing by the cylinder-machine is executed, not only with greater accuracy than by the wooden hand-block as practised in India up to this day, but with an almost incredible saving of time and labour. A single machine, with one man to regulate the engraved pattern-rollers, is capable of printing as many pieces as two hundred men could do with the hand-block in the same time. A length of calico equal to one mile has been printed with four different colours in one hour, or twenty-eight yards in one minute.

It has been estimated that one-seventh of all the cotton spun and manufactured is devoted to printing.

Owing to the great improvements in machinery, and the discovery of new colouring matters, with the modes of fixing them *fast* on the cloth, print-works have now been established all over Europe. The reduction in price of cotton prints within the last fifty years is a striking illustration of the advancement which has been made in Europe in calico printing. England has by far the largest portion of this trade, and especially the export trade. America produces next in quantity to

England.   France and Switzerland next to America in quantity, but far superior in quality.   Austria and Bohemia also produce for their own markets.   Holland, Belgium, Russia, Naples, and even such poor countries as Spain and Portugal, have their print-works, where the most improved machinery is applied to the production of printed goods.

But India, where the art of calico-printing first originated, and which used to export printed goods to Europe—that country has almost lost her occupation. And why?   The art remained in the same state from generation to generation.   With shame it must be told that up to this time the old process of printing is still entirely followed, without the least improvement.   In a country embracing an area of thousands of square miles, with a demand for printed goods, with plants producing beautiful dyes, not a solitary work has been established as yet to print goods with the improved machines—notwithstanding that a moderate capital will suffice for a complete calico print-work.   India is poor even with a rich and fertile soil, and will remain so as long as she depends entirely on agriculture, and (with other requisite changes) improved machinery for manufacturing is not introduced.

CALICO PRINT WORKS

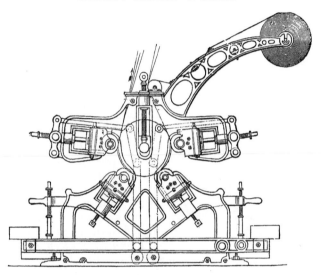

PRACTICAL PROCESSES.

Cylinder-Printing Machine.—The cloth to be printed is first bleached, then wound in rolls by a machine called a condroy, on wooden beams, and brought to the cylinder-printing machine, where the cloth is printed in one, two, three, four, or sometimes as many as twenty colours at the same time. The pattern is engraved on small copper tubes or rollers, which are placed round a large drum of the calico-printing machine. Below each engraved cylinder is a clothed wood roller which partly dips in a trough containing a thickened colour or mordant. As the drum revolves, the colour is rubbed by pressure into the engraved parts of the cylinder. The calico is first carried round the main drum, which is covered with felt, and passes underneath each engraved cylinder, which presses the design on the cloth, as it passes from one roller to the other. The different parts of the pattern falling in their exact places, a proper effect is produced.

The excess of colour is scraped off from each engraved roller by a self-acting sharp knife called a *doctor*, which is so arranged that the scraped colour falls back

again into the colour-box attached to each roller. There is also another arrangement in the machine by which the fibres acquired by the rollers from the calico in the process of printing are removed. A piece of twenty-eight yards long is usually drawn through in about two minutes and conducted to the ageing room.

Engraved Pattern Rollers are now engraved by machinery. For that purpose the roller of brass is first coated with varnish; then put in a machine, where the roller is made to revolve slowly; and the pattern is traced on it by a diamond point. After having been etched on the whole

surface, the roller is placed in dilute nitric acid, which dissolves and deepens the lines exposed by the removal of the varnish—the parts covered by varnish remaining unacted on. There is another method, but not applicable to all descriptions of engraving. The pattern is first enlarged by the camera, and traced on paper in a dark room ; then transferred to a zinc plate, and laid on the curved bed of the pantograph machine.

These engraved rollers weigh from 80 to 120 lbs. each.

Colour-House.—In calico printing, the colouring matter, or the mordant, in order to prevent it from running or extending beyond the proper limits of the

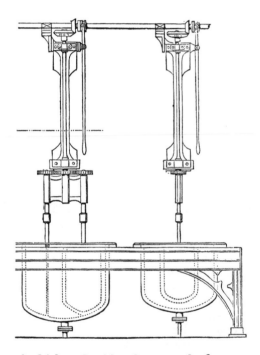

design, is thickened with wheat-starch, flour, or gum. In the colour-house a range of copper pans are fitted with steam and water pipes, and the colours are agitated or stirred by machinery, which was formerly done by manual labour with flat sticks. Only steam-heating being used, any amount of heat required is applied at pleasure by turning on the steam pipes. The process of boiling and cooling is rapid and very certain, and saves a good deal of time and labour.

Ageing Room.—The pieces, after being printed, are dried in the drying machine and then exposed to the action of the air in a large spacious room, kept at an equable temperature, called the *ageing* room. The time of exposure varies according to the style of the work.

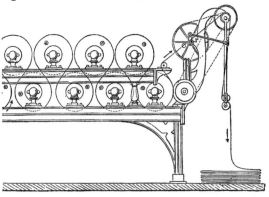

In this room, colours printed on the cloth undergo a change, and become attached to the fibres. In countries where the air in summer is very dry, the ageing room is near or over a reservoir of water, with open boarding for a floor, to admit the watery vapour.

Dunging.—Although in the ageing room the colours combine with the cloth, a quantity of superfluous matter remains deposited. Formerly this was removed by passing the dried cloth through a warm solution of cow dung and water. The superfluous matter combined with the dung, and the cloth was made fit to receive dyes. Dung being very objectionable for obvious reasons, substitutes for it are now used, prepared from bones. In the dung-cisterns, as they are still called (though no dung is used), fifty or sixty pieces are passed per hour, and on leaving, well rinsed in water, then washed in the washing machines. They are now to be exposed to the infusion of the dye stuff.

Dye Becks, or vessels for dyeing, are now constructed of iron for madder work. In the middle of the dye

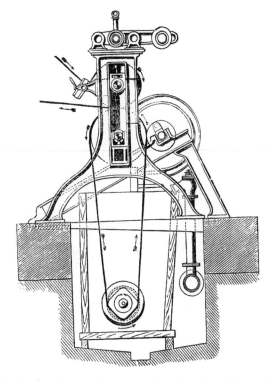

beck is placed a wince or reel; this is set in motion, steam is turned on, by which means the dye stuff solution becomes heated, while the pieces continually revolve with the reel so as to bring each portion successively into the air, agitating the dyeing matters at the same time. When the dyeing is finished, the pieces are put into the washing machines two or three times. With this arrangement five workmen do all the work, which even a hundred hands formerly could not manage to do.

Clearing and Finishing. — Madder goods generally are cleared with soap in a beck similar to the dye beck, then washed and dried. The cloth is then taken to the

finishing room, where, after passing through a starching machine, it is dried by the steam-drying cylinder machine, and made smooth by glazing in the calender, so that it has the appearance of having been ironed.

DIFFERENT STYLES OF CALICO PRINTING.

Madder Style.—The styles of calico printing are numerous, and require somewhat different processes. But the most important and the most extensively practised, forming the bulk of the cotton prints, is the madder style. It derives its name from being chiefly practised with madder or munjeet. Not only does it yield a great number of beautiful shades of colour, but they are all of the utmost degree of permanency,—fast colours, resisting wear, friction, and washing. In this style the thickened mordants are printed on cloth; the cloth is dried, and then conducted to the *ageing* room, where a chemical change takes place; the undecomposed mordant which remains on the cloth is separated, and the thickening paste is removed by passing it through a solution of dung-substitutes, with which it combines. It is then washed in water in the washing-machine, and exposed to the dyeing becks, containing colouring liquor, and washed again in water. The next process is the "clearing," or removing the excess of the colouring matter loosely attached, either by rinsing in clean water, soapy water, or lime water, and then giving a finish to the cloth. An endless variety of tints is obtained from the same madder beck containing the colouring liquor, by mixtures of different mordants in different proportions, applied to the cloth in the printing machine by engraved rollers; and thus, from the same colouring liquor, from red to chocolate, all shades of lilac and purple, up to black, violet, and purple, are obtained.

Indigo or Resist Style.—This is second in importance to the madder style, and indeed hardly second, though its appearance is common and unattractive. In this style a resist paste is printed on white calico, which is then dipped in the indigo vat till the shade of blue wanted is obtained. The resist having the property of preventing the indigo fixing on the printed parts, the result is, dark blue ground with white. *China blue* is a modification of the indigo blue style; the pattern is introduced by indigo colours printed on white cloth; the pieces are next passed to a process which fixes the indigo on the cloth.

Padding Style is applicable for mineral colours. The calico is first passed into a padding machine, which is similar to the starching machine, only the trough contains thickened colour instead of starch. Different coloured figures are afterwards raised by the topical application of other mordants, joined to the action of the dye bath. To produce a design in a mineral colouring matter, the cloth is printed with one saline solution, and afterwards padded uniformly with the other, or otherwise rinsed in the ordinary way.

Discharge or Bandana Style.—The cloth is first dyed, and then placed in a hydrostatic press, covered with a perforated engraved metal plate. Strong pressure being applied by machinery, a solution of lemon juice, or some other substance, is allowed to penetrate below the perforations, and the colour is immediately destroyed. It is in this style of work the celebrated Indian *bandana* handkerchiefs, with white figures on a dark ground, have been imitated, principally in Glasgow. This ingenious process was only invented in 1811, in Europe. The method practised in India (as already indicated) is by binding firmly with thread those spots of the cloth which are to remain white or yellow, while the rest is subjected to dyeing operations.

Topical and Steam Colour Style is largely employed in producing furniture prints, and so called because in this style the application of the thickened colours is applied topically, or mixed with the mordant when any is required, in a state fit to penetrate to the interior of the fibre. After the cloth is dried, the colouring matter is fixed *fast*, either by exposure to air or by the action of steam; and the operations of dunging, dyeing, and clearing are omitted. When the cloth is removed, after the steaming process, it is soft; but on exposure to air for a few seconds only, the thickeners solidify, and are separated by a gentle wash.

ESTIMATE FOR CALICO PRINT WORKS.

Machinery as shown on PLAN G.

	£
Bleach-house machinery, including sewing-machine, singeing-stove, 2 high-pressure keirs, 2 washing-machines, 1 pair of squeezers, 1 drying-machine, 1 condroy, 5 steam-engines, with shafting, pulleys, winces, &c., for carrying cloth, &c. .	1,250
Engraving-room machinery, including 2 engraving-machines, 1 self-acting ruling-machine, 1 polishing-lathe, 1 pair of clams, 1 lathe, 1 steam-engine, with shafting, pulleys, complete .	938
Printing-room machinery, including 1 machine for printing in 1 colour, 1 ditto for 3 colours, 1 ditto for 4 colours, 1 ditto for 6 colours, 4 steam-engines for driving, 4 sets of drying and blanket framing, lapping and blanketing, steam-pipes, valves, &c., complete	1,980
Colour-shop machinery, including a set of double-cased copper colour-pans, with all steam and water pipes, grinding rollers, &c., with fittings complete	450
Madder dye-house machinery, including 1 set of iron dunging-cisterns, with rollers, steam and water pipes, 9 iron madder dye-becks, with winces, peg-rails, &c., 2 washing-machines, 1 pair of squeezers, 1 pair hydro- or water-extractors, 5 steam-engines, pipes and valves, complete	1,665
Finishing-room machinery, including 1 starching-mangle and 1 steam drying-machine, 2 finishing mangles and 1 drying-machine, 3 steam-engines, pipes and valves, complete .	1,068
Steaming-house machinery, including iron folding-doors, wrought-iron carriage, indicator, pipes, valves, &c. . .	93
Mechanics'-shop machinery, including 1 double-geared self-acting lathe, 1 hand-lathe, 1 planing-machine, 1 drilling-machine, 1 grindstone, 1 steam-engine, with shafting, pulleys, &c.	650
2 boilers, for supplying steam to all the works, on the most improved principle, including pipes and valves . . .	750
Accessories and extras, including 100 engraved copper rollers for printing, ready for use, and set of felts for printing-machines, tools, sundry articles, &c.	950
	£9,794

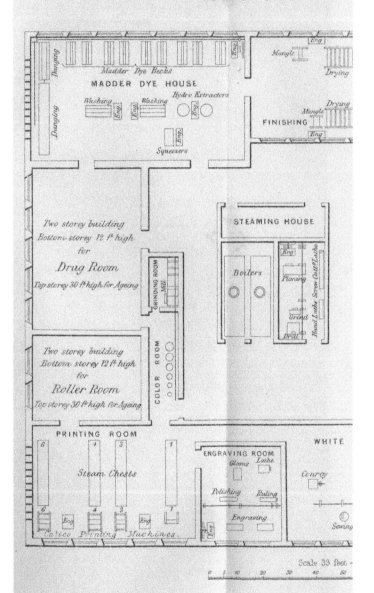

PLAN OF CALICO PRINT

MADDER DYE HOUSE

Madder Dye Becks

Washing · Washing · Hydro Extractors

Squeezers

Mangle · Drying

Drying

Mangle

FINISHING

Two storey building
Bottom storey 12 ft high
for
Drug Room
Top storey 30 ft high for Ageing

STEAMING HOUSE

Boilers

Planing

Grind

Drill

GRINDING ROOM

Mill

Hand Lathe Screw Cut. Lathe

Two storey building
Bottom storey 12 ft high
for
Roller Room
Top storey 30 ft high for Ageing

COLOR ROOM

PRINTING ROOM

6 · 4 · 3 · 1

Steam Chests

6 · Eng · 4 · 3 · Eng · 1

Calico Printing Machines.

ENGRAVING ROOM

Glooms · Lathe

Polishing · Ruling

Engraving

Eng

WHITE

Conroy

Sewing

Scale 33 feet

0 · 5 · 10 · 20 · 30 · 40 · 50

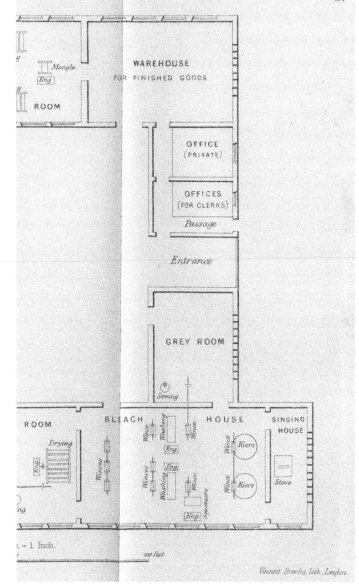

WAREHOUSE
FOR FINISHED GOODS

Mangle

Eng

ROOM

OFFICE
(PRIVATE)

OFFICES
(FOR CLERKS)

Passage

Entrance

GREY ROOM

Sewing

ROOM

BLEACH HOUSE SINGING
HOUSE

Drying

Washing

Winces

Eng

Wince

Winces

Eng

Eng

Wince

Washing

Wince

Squeezers

Eng

Kiers

Kiers

Stove

Wince

Wince

1 Inch.

100 feet

Vincent Brooks, lith. London.

DIMENSIONS OF CALICO PRINT WORKS.

Machinery as shown on PLAN G.

Scale, 33 feet = 1 inch.

	Length, feet.	Breadth, feet.	Square feet.	Height, feet.	Cubic feet.
Grey-room . . .	41	33	1,353	14	18,942
Bleach-house, &c. .	88	45	3,960	14	55,440
White-room . . .	46	45	2,070	14	28,980
Engraving-room .	32	45	1,440	14	20,160
Printing-room . .	63	45	2,835	14	39,690
Roller-room } Ageing	46	32	1,472	42	61,824
Drug-room, &c. }	61	45	2,745	42	115,290
Colour mixing-room	60	18	1,080	12	12,960
Madder dye-room .	81	47	3,807	14	53,298
Finishing-room . .	61	39	2,379	14	33,306
Warehouse . . .	57	39	2,223	14	31,122
Offices	39	33	1,287	14	18,014
Entrance	20	33	660	14	9,240
Steaming-house, &c.	42	51	2,142	14	29,988
Total of square feet and cubic feet			294,53		528,254

REMARKS ON CALICO PRINT WORKS.

MACHINERY AS SHOWN ON PLAN G.

PLAN G shows the arrangement of machinery for calico print works, including also the necessary machinery connected with it for bleaching, dyeing, and finishing about seventy thousand yards of cloth per week.

The whole plan is so arranged that one process will follow another without any unnecessary loss of time or labour. On the right-hand side of the main entrance is a room for grey cloth, opening into the bleach house; after the cloth is bleached, it will be taken to the calico print room, opening into the colour mixing shop, the engraved roller room, and the room for engraving the designs on rollers. The cloth, being printed by the cylinder printing machines, dried, and "aged," will be taken to the madder dye house to be dyed, and then to the finishing room; next to this is the warehouse, which communicates with the offices.

The four calico-printing machines for printing in one, three, four, and six colours respectively at the same time, as well as the other machinery shown on plan, are all driven by small steam-engines, one attached to each machine, which system is preferred and adopted in almost all modern calico print works in England and Scotland. It works more smoothly, it does not require heavy shafting, and the engines are under the direct control of the workman, and very easily regulated.

The engraved copper rollers of patterns generally weigh about 120 lbs each, the price of which varies according to the price of copper and the design to be engraved; but on an average each engraved roller will cost about £7. Machinery for engraving rollers is also included in the estimate, as well as the cost of one hundred engraved rollers, which number will be sufficient for starting the works, as fashions do not change so frequently in India and other parts of Asia as in Europe.

In calico printing, the madder style being the most important and most extensively practised, Plan G has been especially adapted more for that style of printing than any other; while in the arrangement shown on Plan F, in the dye-house department, the machinery is suitable for indigo dyeing, which style is also extensively practised, and is hardly second in importance to the madder style.

It is of great importance that works for bleaching, dyeing, and printing should be erected where pure soft water may be had in abundance. If the water is not pure it imparts a tinge to the goods, and occasions also a loss of dye-stuff. In some very large works in England, where the quantity of water consumed daily amounts to 600,000 gallons, the water is made pure by mixing with it the refuse of the madder dye-pans, which combines with the impurities of the water in an *insoluble* form, and by allowing it to settle in large reservoirs and then passing through beds of gravel, water is obtained clear and fit for use. Rain-water and water from wells is generally better adapted than

I

river water. There are simple methods of rendering hard impure water well adapted for dyeing purposes.

The bleaching necessary for printing calicoes is more complete than that which will suffice for calicoes intended to be sold in the white state; because the whiter the cloth is, the more light it reflects from the surface, and the colours of the dyes appear more brilliant. The different styles of calico printing have been described already in a former page.

In consequence of the great increase in price of gum arabic, arising from its extensive employment in calico printing, a great variety of artificial gums are now prepared from sago, rice, and wheaten flour, for thickening the colours, or mordants, used in printing. Wheaten flour, when heated, generally bears the name of British gum.

The cost of the calico printing machinery, complete, as shown on plan, including freight, will amount to about £10,000. The cost of the buildings, &c. may be reckoned from the dimensions of the plan, which consists of four wings, lighted from side windows—with the boiler, steaming house, and mechanics' shop placed separately in the yard by itself. The total cost of the whole print-works complete, ready for working, may be safely calculated under £25,000.

JUTE MANUFACTURES.

INTRODUCTION.

JUTE is the material of which gunny bags and gunny cloth, bagging for cotton, carpets, and such other fabrics, are made. In India, jute manufactures must be placed next in importance to cotton, on account of the large consumption both in India and in other foreign countries, and also because the demand will probably continue in every market of the world.

The jute plant is most extensively cultivated in Bengal. Its culture is easy, the growth rapid, and the production comparatively large. The crop, when ripe, is cut down close to the roots, and after being steeped in water for a week or longer, the bark separates easily, and the silky fibre is detached, cleaned, assorted, and if intended for export to Europe, is pressed into bales containing 300 lbs. each. The annual production of jute in India is now more than 300,000 tons.

Various coarse fabrics are made from jute in Bengal, known as *gunnies* and *gunny bagging*, of which whole cargoes, amounting to tens of thousands of pieces, are annually taken by the Americans as a cotton-bale material, in lengths of about 30 yards, weighing about 6 pounds each. As it takes about 6 yards for wrapping every bale of cotton, a crop of 3,000,000 bales will require 18,000,000 yards of bagging, and one-third of this supply is exported to America from Bengal. In some years as much as 270,000 tons of manufactured bagging have been shipped from India to foreign countries.

In Bengal, jute has hitherto been worked by very rude processes, and made into cloth, and has formed the grand domestic manufacture for hundreds of years. In Great Britain it has only been known within the last quarter of a century. Machinery worked by steam has been applied, as in cotton goods, to spin and weave jute, whereby the production has been largely increased. Owing to the raw material being very cheap, combined with a considerable amount of spinning quality, and the smooth appearance of jute cloth made by improved machinery, this manufacture has made rapid progress in Great Britain, continues daily to increase, and has been extended in a short period to an extraordinary degree. Colossal jute factories have been erected in Great Britain, in some of which the consumption of jute exceeds 500 bales every week, and in two or three factories 1,000 bales per week. The various descriptions of jute goods manufactured in these factories are baggings, sackings, hessians, sheetings, ducks, carpeting, &c. The major portion of the jute cloth is made of yarn in the green or natural state; but in many sorts the yarn is bleached, dyed, and finished. Besides fabrics composed wholly of jute, many sorts of unions are made by incorporating with jute, cotton, woollen, flax, and tow yarns. The jute yarns used for carpets are of varied colours, and are sometimes used in conjunction with cocoa fibre. Even the Brussels, or velvet, carpet is imitated with success in appearance, but not in durability.

For many years all the jute consumed in Great Britain was worked up in Dundee, but the manufacture is now extending in Glasgow, Manchester, London, and other cities. In Dundee alone nearly 50,000 tons are manufactured into bags and sacks annually, the raw material of which is exclusively supplied by India.

In France, jute spinning and weaving are also carried on, where the consumption amounts to about 10,000 tons. In the United States this manufacture has been progressive, notwithstanding that labour there is even dearer than in England.

In the mother country of jute—Bengal—where it has been manufactured for hundreds of years even before its very name was known in Great Britain, no progress

has been made. There are about half a dozen jute
factories near Calcutta, for spinning and weaving jute
by improved English machinery ; but these factories
have been erected only recently, by the enterprising
Britons, thousands of miles from their homes. The
honour of introducing improved machinery for jute is
due to them.

The establishment of steam jute factories in Calcutta,
remarks a Dundee manufacturer, has not hitherto had
an injurious effect upon the trade in Dundee. The
goods they manufacture in Calcutta and its neighbourhood, have come more into competition with the production of the native looms, as they have been chiefly
used for bags for rice and other produce of the country,
which was formerly shipped in gunny cloth made by
hand. The consumption of cloth for bagging is immense, and constantly on the increase. The manufacturing establishments at Calcutta are reported to have
orders always in advance to last them for months, and
have yielded a handsome return to the proprietors. In
Bombay, where bagging is so largely consumed for
cotton bales, linseed, and other produce, the *first* jute
factory is expected to work in a short time.

Jute, sunn, hemp, and flax, are spun on machinery
on the same principle, but with differences in size and
proportion, and in other minor details. Flax fibres contain a large quantity of gum, and there is a great
resistance in spinning it. The fibres of hemp are inferior to those of flax, and more difficult to spin ; while
jute, besides being very cheap, is also freely spun, and
this accounts for the very rapid progress of the jute
manufacture in Great Britain, compared with linen or
hemp.

JUTE.

MANUFACTURING PROCESSES.

(1.) *Oiling.*—In jute factories in England, after cutting open the closely-packed jute from the bales, it is spread out; oil and water are sprinkled on it, and allowed to remain in the batch for two days. The application of oil softens the material, makes it more pliable, and gives it a better spinning property. Then, in some factories, it is taken to a softening machine, and pressed between rollers heavily weighted; but in several factories the softening machine is not used, the jute being taken direct to the breaker-card machine.

(2.) *Breaker and Finisher Cards.*—The jute is first passed through the breaker-cards, consisting of rollers with teeth, the main cylinder of which is four feet in

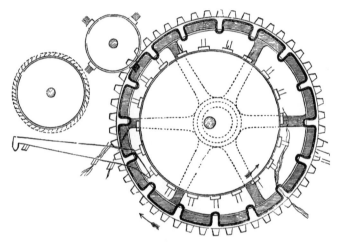

diameter. The teeth strike the jute and tear it up, and a silver or ribbon of jute is formed. The finisher-card continues the same operation as the breaker-card, but the teeth of the card-clothing are finer and more closely set. The slivers of jute are led off at the side of the finisher-card, where they are deposited in cans.

(3.) *Drawing Frame* is similar in principle to that used in the cotton manufacture. In this machine a number of jute slivers are drawn out repeatedly, which increases the length and diminishes the thickness of each sliver. There are the first and second drawing-frames, which are in principle the same. The object of the second set is still to reduce the thickness. At one end of the machine is a can full of slivers brought from the finisher-card. By the drawing-rollers the sliver of jute is taken into the machine, and after passing under other rollers moving at different speeds, the sliver falls into another revolving can opposite.

(4.) *Roving Frame.*—This machine is for the purpose of giving a slight twist, and of winding jute slivers upon bobbins, preparatory to spinning. The improvements lately introduced in cotton roving-frames for giving a permanent and independent support to each

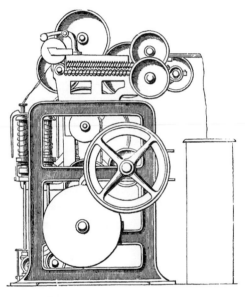

spindle, and in maintaining a perfect steadiness of action at a much higher velocity, have also been applied to jute roving-frames. The can of sliver is brought from the second drawing-frame to the roving-frame, where it is twisted and wound upon wooden bobbins.

(5.) *Dry Spinning Frames.*—On these frames jute yarns are spun into either weft or warp for weaving. It is called *dry* spinning, because in spinning-frames

made on the same principle for linen, water is used, as
flax yarns require to be wetted in order to render them
more pliable and easy to twist.   Water is contained in
a trough, which is supplied to the roller by the capillary
attraction of a piece of cloth immersed therein.

The frames for spinning jute very much resemble

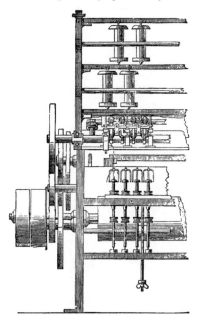

the throstle used in cotton spinning. The speed is from
3,000 to 4,000 revolutions per minute.   When the
operation of spinning is finished, the thread is perfectly
formed.

(6, 7, 8.) *Winding, Beaming, and Pirning Machines.*
—The winding-machine is for winding the warp yarns
on large bobbins ; the beaming-machine to wind the
yarn on beams for the purpose of placing it on looms
for weaving ; and the pirning-machine to make cops or

pirns for the shuttles. All this preparatory machinery is required for the purpose of weaving the spun yarns in the looms. The shuttle containing the weft yarn is much larger and stronger than that used in cotton looms.

(9.) *Looms.*—The jute-looms for weaving sackings, gunny bags, and other coarse fabrics, plain or twilled,

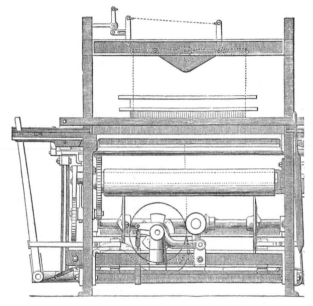

are much stronger and heavier than calico looms. There is a contrivance in the loom for making a proper shed for the very large shuttle used, and also for reducing the chance of breaking the warp threads. There are looms made with more than one shuttle, for throwing variously-coloured wefts for jute carpets, and such other fabrics, which are also manufactured extensively in Dundee. The loom for making sail-cloth is similar in principle to that of jute.

ESTIMATE FOR A JUTE FACTORY.

Machinery as shown on PLAN H.

	£
2 Shell-breaker cards, cylinders 4 by 6 feet, 48″ on wire, complete, with all fittings, and card clothing	523
4 Finisher-cards, cylinders 4 by 6 feet, 48″ on wire, complete, with card clothing	1,344
4 First drawing-frames, 2 heads of 4 bosses each, with patent lowering motions	360
4 Second drawing-frames, 2 heads of 4 bosses each, complete .	360
4 Regulating roving-frames, 48 spindles each, complete . .	1,380
10 Double dry spinning-frames, 128 spindles each, complete .	2,730
2 Shell-breaker cards, cylinders 4 by 6 feet, 48″ on wire, with card clothing, complete	523
4 Finisher-cards, cylinders 4 by 6 feet, 48″ on wire, complete .	1,344
4 First drawing-frames, 3 heads of 4 bosses each, complete .	523
4 Second drawing-frames, 2 heads of 4 bosses each, complete .	425
4 Regulating rotary-frames, 48 spindles each, complete . .	1,175
6 Double dry spinning-frames, 108 spindles each, complete .	1,518
5 Winding-machines, 3 beaming-machines, 400 bobbins each; 4 pirning-machines, 60 spindles each, to make cops, complete	1,202
50 Patent looms for heavy sacking, 36″ reed space, with positive taking-up motion, yarn beams, flanges, complete . .	1,335
50 Patent looms for gunny bagging, very strong, and similar to the above, complete	1,450
If required for *twilled* sacking and bagging, add for 100 looms	250
Accessories and extras, consisting of 50 yarn beams, 12 beam stands, 4 drawing-in frames, 300 shuttles, 100 gross bobbins, 100 gross pirn heads, 150 pair pickers, &c. .	540
2 Steam-engines, both condensing, one with cylinder 33″, stroke 5 feet, and one 16¼″, stroke 2 feet 8 inches, with fly-wheels, governors, condensers, air-pumps, and all fittings, complete.—2 wrought-iron boilers, 25′ × 6′ 6″, with 2 flues, with all mountings, complete.—Gearing, shafting, pulleys, fixings, &c., complete	3,700
	£20,682

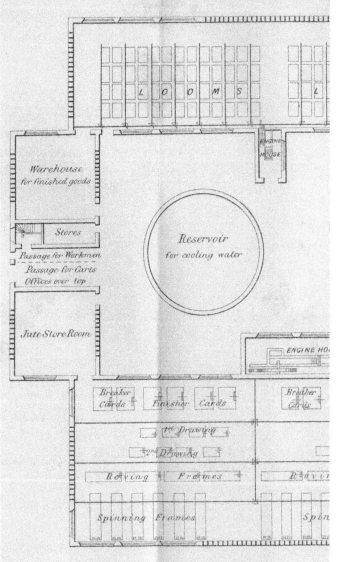

L O O M S   L

Warehouse
for finished goods

Stores

Passage for Workmen
Passage for Carts
Offices over top

Jute Store Room

ENGINE
HOUSE

Reservoir
for cooling water

ENGINE HO

Breaker
Cards   Finisher Cards   Breaker
Cards

1st Drawing

2nd Drawing

Roving Frames   Roving

Spinning Frames   Spin

Scale 48 feet = 1 Inch
0 5 10   20   30   40   50

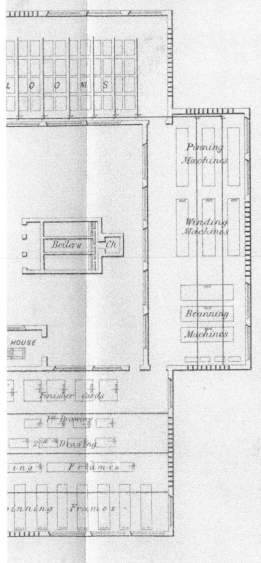

DIMENSIONS OF A JUTE FACTORY.

Machinery as shown on PLAN H.

Scale, 48 feet = 1 inch.

	Length, feet.	Breadth, feet.	Square feet.	Height, feet.	Cubic feet.
Front wing, containing store-room for raw material, warehouse, entrance .	124	46	5,704	12	68,448
Offices over top story	50	46	2,300	10	23,000
Side wing, containing preparation and spinning machinery	211	84	17,724	12	212,688
Intermediate wing, containing preparation for weaving	129	55	7,095	12	85,140
Side wing, containing 100 looms . .	211	56	11,816	12	141,792
Engine-house for spinning . . .	58	20	1,160	14	16,240
Engine-house for weaving . . .	27	16	432	14	6,048
Boiler-house. . .	40	27	1,080	10	10,800
Total of square feet and cubic feet			47,311		564,156

REMARKS ON JUTE FACTORY.

THE estimate shows the cost of machinery for a jute factory of moderate size, consisting of about 2,000 spindles and 100 looms. Near Calcutta three jute factories have been in active operation for several years past. The Borneo Jute Company, with the extension they have made, is working 3,500 spindles. The first jute factory in Bombay, called George the Third Jute Spinning Company, owing to the monetary crisis of 1866, went into liquidation even before it commenced to work.

The spinning machinery as shown on plan is arranged in two distinct divisions for weft and warp to spin yarns 7 to 24 lbs, and is quite ample to supply yarns for the looms. If this arrangement of machinery in two divisions be not approved, that one adopted for the cotton mill machinery may be easily substituted in its place.

The manner of counting jute yarns is quite different from that of cotton. The heavier yarns are the higher numbers, just the reverse of cotton yarns.

The weaving machinery, consisting of 100 looms, is adapted for weaving gunny bags and sackings. For twilled sacking, twilling motions are required in looms, for which an extra charge is made, included in the estimate.

There are two separate steam-engines provided, one

for driving the spinning machinery, and the other for driving the looms, which arrangement is in some respects preferable; however, it will be a simple thing to drive all the machinery by one engine if desired. As the total amount of horse-power required will remain the same, it matters little if the machinery is driven by one engine or two.

The production from 100 looms will be according to the fabric to be produced; the coarser the fabric, the larger the production will be.

The buildings for the factory consist of four wings, connected with each other. The machinery is all arranged on the ground floor. Over the entrance for the passage of goods and the store room is an upper story for offices, which will give a pretty effect to the entire building. The plan is so arranged that in case of a small extension there will be room for additional machinery, without increasing the size of the buildings, which, if not desired, can be proportionately diminished in the first instance.

The cost of machinery amounts to somewhat above £20,000. The total cost, including the necessary buildings and erection of machinery in complete working order, may be put down in round figures at £40,000.

The accounts of the Calcutta jute factories have not been published; they are kept secret for trade purposes. However, from the extensions they have made to their works, and from other reasons, there is no doubt that these companies have declared handsome dividends.

QUANTITY AND VALUE OF JUTE AND JUTE BAGGING EXPORTED FROM INDIA,

For each of the Years ending April 30th.*

Year.	Jute.	Value of Jute.	Bagging: value.
	Cwts.	£	£
1866	1,638,174	1,499,533	
1865	2,120,813	1,793,029	102,858
1864	2,513,887	1,507,035	111,207
1863	1,266,884	750,456	131,628
1862	1,232,279	537,610	186,845
1861	1,092,668	761,201	359,343
1860	761,201	290,018	333,977
1859	317,890	525,090	392,424
1858	788,820	303,292	217,194
1857	673,416	274,957	376,252
1856	882,715	329,076	302,388
1855	699,566	229,241	215,335
1854	509,507	214,768	174,790
1853	349,797	112,617	231,159
1852	535,027	180,976	287,411
1851	584,461	196,936	160,397

* From Statistical Abstract presented to Parliament, 1867.

QUANTITY AND VALUE OF JUTE BAGS EXPORTED FROM GREAT BRITAIN.

Year.	Quantity.	Value.
	Dozens.	£
1866	1,338,184	675,740
1865	1,137,862	696,291
1864	971,871	749,422
1863	894,436	555,282
1862	802,095	388,724
1861	642,848	307,583
1860	623,752	293,593

SILK MANUFACTURES.

INTRODUCTION.

THERE are few things more
wonderful than the mighty
results frequently brought
about by very small and apparently inadequate means.
Few objects can appear more
insignificant than the silkworm; yet, by the united
labours of millions of these
creatures, the raw silk is
made, the trade in which
amounts every year to millions of pounds sterling,
and enables man to weave
beautiful fabrics from it.

In China, even more than
two thousand years before
the Christian era, this valuable insect was made serviceable to man. From that
country the art of rear-
ing the silkworm passed
into India and Persia; and
up to this day China, India,
and other parts of Asia
supply to Great Britain and
other European countries
the bulk of the raw material.

K

The silk manufacture is now a source of wealth to European nations, though it was only at the beginning of the sixteenth century that the silkworm was introduced into Europe. Two priests brought some eggs from China to Constantinople, from which place the art of rearing the silkworm spread into Greece, Italy, and France. The mulberry tree, on the leaves of which the silkworm is fed, and which is a native of China, was first planted in France in the year 1564, and thence propagated into other parts. France, so famous now for the excellence and beauty of its silk goods, produces a large quantity of raw silk, but it consumes all it produces, and the deficiency is made up by importations from China and India.

The silkworm is subject to many changes. The egg, which is smaller than a grain of mustard seed, sends forth a caterpillar, which as it enlarges casts its skin three or four times. In the course of about a month it acquires its full size, leaves off eating, and begins to discharge a secretion from its head, like a common spider. This secretion hardens into silken lines, which are coiled instinctively into a nest round the worm itself, and which is called a cocoon.

The caterpillar remains working within the cocoon, spinning its beautiful threads till its body diminishes one half. The cocoon being complete, the worm transforms, and after a week emerges in the form of a perfect winged insect, the silk-moth. In order to escape the cocoon, it moistens the interior with a liquid, dissolves the gum that holds the silken lines together, and escapes. At this time the male and female moths couple together. The insects have a short life and only one object to accomplish — a continuation of the species. The female moth lays her egg in two or  three days and then dies. The eggs can be preserved in a low temperature. If the heat advances rapidly at an unusual season, it is not allowed to act on the eggs, otherwise the caterpillars become hatched before new leaves appear on the mulberry trees to feed them. In France, in large establishments, the eggs are placed in large apartments, and where the heat is every day gradually increased till it reaches a certain point; then the eggs are hatched together in large broods, and nature completes her mysterious work in a short time. The newly-hatched worms feed on tender mulberry leaves strewed on a piece of paper.

In the French settlements in India, experimental establishments for rearing improved species of silkworm have been erected; eggs and cocoons have been transmitted from India, Persia, and China to France, and the worms have fed well on the oak leaves of the forests. The French Government have from time to time also offered large grants of money for improving the race of silkworms. By improved plans they have succeeded in obtaining a kind which yield large and equal-sized cocoons of a pure *white* colour, the silk of which is equal in all its length, strong and shining. From the ordinary species about eight pounds of reeled raw silk is obtained from one hundred pounds of cocoons. In England, the rearing of the silkworm has been tried, but on account of the cold winds it has not succeeded commercially.

There are several species of silkworm known in India. No less than thirty are fully described by Mr. F. Moore, in his "Synopsis of Asiatic Silk-Producing Moths." The Tusseh silkworm is the most common species in Bengal. Millions of its cocoons are annually collected. The fabric woven in Bengal, chiefly at Midnapore, from this silk, is called Tassar cloth, and is of very strong texture. The female cocoons are heavier and rounder than the males, and a due proportion of the finest cocoons of each sort is preserved for the production of eggs. The worms feed chiefly on the leaves of the castor-oil plant. The Chinese and also the European cocoons are much larger than those of India. Mr. Bashford, of Sirdah, attempted to enlarge the native cocoons by crossing the Indian silk-moths with the European varieties. He obtained several mixed breeds, much larger and stronger than the ordinary Bengal stock, producing beautiful cocoons. The best silk in the world is that obtained from the Chinese cocoons. China still sends to Great Britain its unfailing supply of *Tsaltee*. Bengal still furnishes a large amount of raw silk, and also silk in its manufactured state, but of inferior quality to that of China.

The principal crop of cocoons in India is assorted according to quality. The life of the worm enclosed in the cocoon is destroyed by heat, and the floss silk over the cocoon is removed. After being placed in hot water, the gum softens, and a number of loose ends are reeled from the cocoons together. In France this is done by improved mechanism in a very ingenious manner, without any waste. The reeled silk is made up into hanks and bundles for the market. Its form and quality differ in the several countries. The colour

in the native state is generally a golden yellow, but some varieties of silk are perfectly white. A thread of silk is three times stronger than one of flax, and twice stronger than hemp of the same diameter; in fact, it is the strongest of all the textile fibres. The production of raw silk is an important branch of industry, open for future skill and exertions in many parts of India; and

the duty of the English Government is to go on en
couraging it by every possible means, as they are doing
in France.

At one time a disease, mysterious as the cholera,
having destroyed myriads of worms in the silk-producing
countries of Europe, millions of eggs were exported
from Asia to Europe, where by crossing the enfeebled
breeds a supply of silk larger in quantity and better in
quality was obtained.

The manufacture of silk goods was commenced in
Europe only in the sixteenth century. For several
years everything was done by hand, as it is up to the
present time in India and China. It was in the year
1716 that the first silk mill moved by water power was
erected in England. A large number of skilful French
weavers settled in England during the persecutions in
France, and the silk manufacture made more progress.
So late as 1838, at Coventry, the highest authorities in
the trade decided that a good silk ribbon could never be
made by steam so well as by hand. But events proved
that even the highest authorities had to learn the direction in which their true interest lay. In the twenty
years between 1838 and 1858, the steam factory system
developed itself with continually increasing rapidity
and energy ; and the application of steam, first pro-
nounced impossible and unprofitable by practical men,
was found to be an inevitable *necessity* even in ribbon
manufacture. One manufacturer after another adopted
the factory system, and erected larger works and more
effective machinery. Silk factories were established at
Coventry, Spitalfields, Macclesfield, Manchester, and in
other towns. There are now more than five hundred
silk factories in Great Britain, in which a variety of

goods to suit the tastes and pecuniary means of nearly all grades of civilised and semi-civilised people are manufactured. In these factories more than 5,000 horse-power is employed, and a capital of about £50,000,000 sterling invested, giving employment to nearly 1,000,000 persons.

In France a large capital has been invested in the silk industry, and its manufactures have attained such perfection that their superiority in taste and execution has not been even disputed by the English. The French have one advantage over the English, that they produce on their own soil a large portion of the raw material.

China and India produce beautiful silk goods, but by slow processes and by manual labour. What gives advantage to the European manufacturer is the application of steam-power and improved machinery for spinning and weaving, whereby economy of labour and expense is effected, rendering the article cheaper and more accessible.

SILK MANUFACTURES.

MANUFACTURING PROCESSES.

(1.) *Winding Machine.*—The hanks of raw silk, as made up in Bengal and China, direct from the cocoons, in continuous lengths, are stretched in the English silk factories upon six-sided reels, called swifts, by which the silk is transferred from the hanks to bobbins. Should the silk in winding be entangled, the bobbins are stopped without breaking the thread, till the silk has been examined and rectified. The winding machines for China and Bengal silk differ somewhat in size.

(2.) *Clearing Machine.*—The bobbins from the winding machine are removed to this machine, the object of which is to remedy the irregularities and knots in the thickness of silk by passing it between the edges of a steel clearer. Each thread is passed from its bobbin through the clearer to another bobbin. Sometimes this

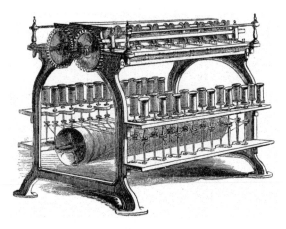

cleaning is affected by passing the silk threads between two steel rollers, adjustable by screws. After the silk has undergone the process of cleaning, it is ready for the spinning machine.

(3.) *Spinning Machine.*—This twists the single thread so as to give it strength and elasticity. It consists of upright spindles on each side of the machine, arranged in two or three tiers; but the three-tier machine being very high, is inconvenient to work. The bobbins of silk from the clearing machine are placed on the upright spindles in such a way that, whilst each spindle is rapidly rotated, the thread is being drawn off and again distributed upon another bobbin placed horizontally. These machines have the usual arrangements of change wheels to put the required twist per inch into the thread.

(4.) *Doubling Machine.*—This winds and lays evenly

together two or more threads of silk from the spinning machine upon a single bobbin. An ingenious

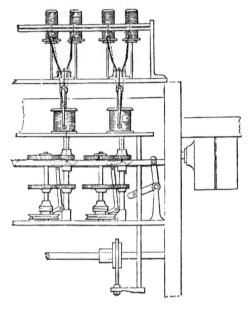

device is employed to stop the winding-on, the moment
one of the threads happens to break. The girl who
attends to the machine pieces-up the broken thread,
and sets the machine in motion again. The ends of
silk from the bobbins of the spinning machine are,
according to the sort of work required, wound in this
machine in one cord or strand, and it is essential that
this strand be laid evenly, with an equable amount of
tension.

(5.) *Throwing Machine.*—On this the doubled threads
are twisted together, their bobbins revolving on upright
spindles as in the spinning machine, and making from
3,000 to 4,000 revolutions per minute. The spun threads

are wound on reels into hanks ready for the dyer. As each reel is filled, it gives notice by striking a bell.

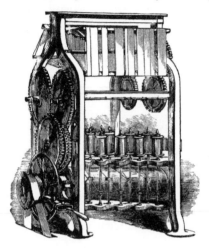

Silk is sometimes thrown by means of the spinning machine only. Where all the operations of spinning silk are carried on, the place is sometimes called a *throwing* mill.

(6.) *Glossing Machine* is designed to stretch the dyed silk, and gives a gloss to it, which was formerly done by hand. The silk yarn is confined in a steam-tight case. By the direct action of steam, the silk is thoroughly saturated, when it is stretched by the pressure applied to the piston. The machine is made especially for silk, which can be stretched a tenth of its length.

(7.) *Soft or Dyed Silk Winding Machine.*—This is for the purpose of winding the silk that has been dyed. The silk thread is wound from the dyed hank on bobbins making 200 turns per minute, in readiness for warping or for putting into shuttle-pirns for weaving in the looms.

(8.) *Jacquard Loom* for facing goods has now entirely superseded every other for producing figured work in the loom. The Jacquard apparatus, named from a Frenchman who invented it, is an addition to the power loom, and is a very ingenious mechanism for raising certain threads of the warp in a certain predetermined order, so that on throwing the shuttles, each

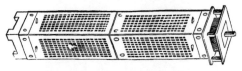

containing a different colour in a certain order, a figured pattern shall be produced. Since Jacquard invented it, a good many improvements in detail have been made, which have been found of essential service in certain classes of weaving. Change in design or pattern is made without any difficulty.

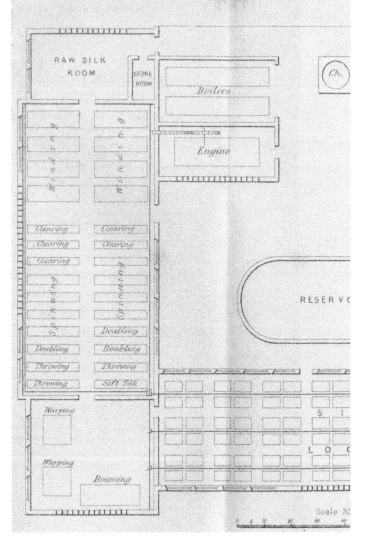

RAW SILK ROOM

STORE ROOM

Ch.

Boilers

Engine

Winding

Winding

Clearing

Clearing

Clearing

Clearing

Clearing

Spinning

Spinning

Doubling

Doubling

Doubling

Throwing

Throwing

Throwing

Soft Silk

RESERVO

Warping

S

Warping

L O

Beaming

Scale 32

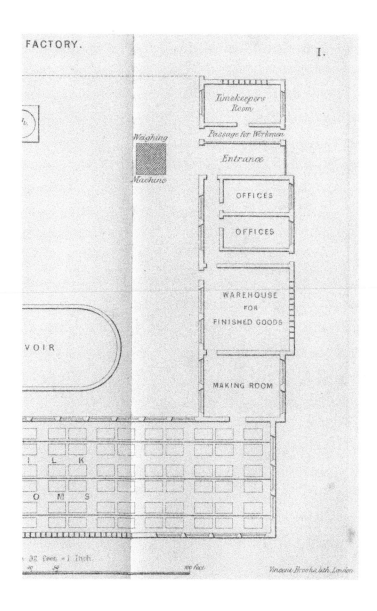

Timekeepers
Room

Passage for Workmen

Entrance

Weighing

Machine

OFFICES

OFFICES

WAREHOUSE
FOR
FINISHED GOODS

VOIR

MAKING ROOM

I L K

O M S

32 feet = 1 inch.

40   50

100 feet

Vincent Brooks, lith. London

ESTIMATE FOR A SILK FACTORY.

Machinery as shown on PLAN I.

	£
8 Winding machines, 608 spindles, pitch 6½″, complete . .	182
5 Clearing machines, 380 spindles, pitch 6″, complete . .	152
8 Spinning machines, 1,536 spindles, pitch 5″, complete . .	485
3 Doubling machines, 328 spindles, pitch 5″, complete . .	97
3 Throwing machines, 576 spindles, pitch 5″, complete . .	207
1 Reel for dyed silk; 1 steam stretching and polishing machine	110
2 Warping mills; 1 bearing machine, complete . . .	76
102 Looms, 36 inches reed space, complete; 102 Jacquard machines, complete	1,316
1 Horizontal steam-engine, 25 horse-power, boilers, shafting, gearing, &c., extras, packing and free delivery on board .	1,460
	£4,085

DIMENSIONS OF A SILK FACTORY.

Machinery as shown on PLAN I.

Scale : 32 feet = 1 inch.

	Length, feet.	Breadth. feet.	Square feet.	Height, feet.	Cubic feet.
Raw silk room	50	27	1,350	12	16,200
Spinning-room	105	53	5,565	12	66,780
Preparation weaving . .	42	50	2,100	12	25,200
Weaving looms	162	44	7,128	12	85,536
Making-up room . . .	24	33	792	12	9,504
Warehouse	31	36	1,116	12	13,392
Offices, &c.	32	36	1,152	12	13,824
Timekeeper's room . .	16	33	528	12	6,336
Engine and boiler-house	45	45	2,025	10	20,250
Total of square feet and cubic feet			21,756		257,022

REMARKS ON SILK FACTORIES.

Silk machinery is very simple. As the silk thread is made by the silkworm, no preparatory machinery is required as for cotton, jute, &c.

Throwing mill means a factory where silk is prepared for weaving by converting it into three forms, known as singles, tram, and organzine. *Singles* is formed by simply twisting the filaments of raw silk to give firmness to the texture. *Tram* is usually made from inferior silk, by twisting together in one direction two or more threads.

Weft for common and flowered silks, velvets, &c., is formed of *tram*, by slightly twisting together in one direction two or more filaments of singles, usually made from common silk, which spreads in weaving. *Warp* for silk goods is made from *organzine*, or silk cleaned, spun, doubled, thrown, and considerably twisted, so as to resemble the strand of a rope. It is made by twisting together two or more singles or tram in such a manner that the twist of the combined thread shall be in the opposite direction to the twist given before, which prevents it untwisting. As organzine is twisted hard, it is used as *warp* for best silk stuffs.

Hard silk is that in which the natural gum is left in

it. *Soft* silk means that the gum has been removed by scouring. *Floss* silk consists of shorter broken threads, waste, &c. It is cut, carded, and spun in England like cotton, for inferior silk goods.

The quality of silk is determined by first winding a certain quantity round a drum, and then weighing it. The weight is expressed in grains, twenty-four of which constitute *one denier*. Twenty-four deniers make one ounce, and sixteen ounces one pound.

Raw silk absorbs moisture, and is increased ten per cent. in weight thereby. This property leads to fraud, which, however, may be detected by enclosing a certain weighted portion of suspected silk in a wire-cloth cage, and exposing it to heat for a certain number of hours with a current of air. The loss in weight shows the amount of fraud. At Lyons, in France, a place famous for its silk factories, there is an office established by Government where this assay of silk is made. The law of France requires that all the silk tried by this office must be worked up into cloth in that country.

The Plan I of Silk Factory shows arrangements for 1,500 spinning and 560 throwing spindles, with 100 Jacquard looms for weaving *figured silks*. Gauze-weaving and satin-weaving can be performed on a loom for plain weaving; but for figured silks, a Jacquard loom will be necessary. A Jacquard can be applied to any plain loom, and costs about £10. Machines have been devised to do all the operations of silk-spinning in one; but different speeds being necessary, there is a great waste and also increased cost in wages; so the plan most commonly adopted is to have distinct and separate machines for each operation required in a silk factory. On plan, in the spinning-room, two rows of machines are shown, driven by two shafts, but one row would be preferable if there is room.

Steam is more advantageously employed in the manufacture of coarse and plain goods than in that of very fine goods. The motive power required for silk machinery is much less than for cotton, jute, or other textile fibres. A steam-engine of ten nominal horse-power will drive all the spinning and preparation machinery, as shown on plan, and an engine of fifteen horse-power will drive 100 looms.

The estimate for the silk machinery as shown on Plan I, amounts to £4,000. The total cost of the silk factory, including buildings, fitted up complete, may be put down at about £9,500. It will be reduced by half if the number of looms be limited to fifty, or otherwise proportionately less by having fifty Jacquard looms, and fifty ordinary looms for plain weaving, instead of 100 Jacquard looms as provided in the estimate. In some places it may be convenient to spin the silk *direct* from the cocoons in the factory, in which case some of the machines specified will not be required.

EXPORT OF SILK FROM INDIA.*

Year.	Raw Silk.	Value.	Goods—value.
	lbs.	£	£
1865		1,165,901	106,612
1864	1,369,556	954,649	115,465
1863	1,228,684	822,892	165,136
1862	1,101,844	686,083	168,806
1861	1,955,656	1,036,728	134,831
1860	1,670,953	817,853	191,509
1859	1,217,438	725,655	213,108
1858	1,580,463	766,673	158,224
1857	1,756,778	782,140	281,450
1856	1,521,765	707,706	341,035
1855	1,195,310	500,105	263,453
1854	1,693,125	640,451	326,571
1853	1,411,536	667,546	315,195
1852	1,469,291	688,640	265,384
1851	1,325,901	619,319	355,223

(Silk goods, manufactures of India, consist of bandanas, corahs, chopas, Tussore cloths, romals, and taffeties.)

RAW SILK IMPORTED INTO GREAT BRITAIN FROM INDIA AND CHINA.†

Year.	Overland, via Egypt.‡	From China.	From India.
	lbs.	lbs.	lbs.
1866	3,405,898	108,201	123,561
1865	5,054,354	136,652	183,224
1864	3,401,116	461,357	167,774
1863	4,779,543	1,695,882	208,029
1862	5,434,785	3,267,044	469,985
1861	4,224,565	2,348,578	136,505
1860	6,768,601	2,185,742	60,985
1859	5,289,773	3,055,262	331,488
1858	3,652,617	2,011,186 ⎫	
1857	4,678,415	6,664,532 ⎬	included in overland
1856	3,124,778	3,722,693 ⎭	

* From Statistical Abstracts presented to Parliament, 1867.
† From Board of Trade Returns.
‡ Overland via Egypt, in transport from India, China, and Japan.

SUGAR.

INTRODUCTION.

SUGAR is consumed throughout the whole world in some form or other. In Great Britain the revenue derived by duties imposed on this one single article amounts every year to more than £5,000,000. Sugar has grown into extensive demand, and the imports into the United Kingdom from India, Mauritius, China, Java, and other sugar-growing countries, are enormous. The importance of sugar, therefore, cannot be too highly estimated, both as regards its culture and production in India and other countries, and its manufacture for local consumption, and for export to Europe and America.

To the inhabitants of India and China, sugar has

been known from a very early period. From India, the sugar-cane was introduced into Arabia, Egypt, and other Western parts of Asia, and then into Europe, where at first it was used for medicinal purposes alone for many years. A physician described sugar as Indian salt, and recommended that a piece be kept in the mouth during fevers. By the Greeks and Romans it was called by analogous names; and they obtained it at enormous cost from India. In Spain, it was introduced by the Moors. The Portuguese planted the sugar-cane in America, soon after that continent was discovered. From Brazil it was transmitted to Barbadoes, from which island at present large quantities are shipped to Great Britain every year. Thus the culture of the sugar-cane spread from one country to another adapted for its cultivation.

When sugar became employed in Europe in sweetening food and beverages, its consumption was inconsiderable, being made use of only in the houses of the rich and great; but with the introduction of tea, coffee, and chocolate, sugar came into general demand.

There are several sugar-producing districts in the Bengal and Madras Presidencies, and the export of sugar forms an important item of commerce with other countries. In India, sugar is also obtained from the date palm-juice. There is a caste in India who make sugar from the kital-tree, which is called *jaggery*. There are several sorts of cane cultivated in India, producing good sugar. The China cane is very hardy, and gives out as many as thirty shoots. Penang, Singapore, and the Straits settlements, produce excellent cane, yielding large quantities of sugar per acre.

Beetroot sugar is largely manufactured in France,

and other countries of Europe, but not in England.
The history of this industry is important, as pointing
out one of the means by which the resources of a
country may be developed, and showing how science,
when applied successfully, promotes manufactures and
commerce. At the commencement of the present century,
Napoleon Buonaparte, during the wars, closed all the
ports, and excluded cane-sugar produced in the English
colonies.   He demanded of his learned men the means
to produce sugar on French soil.   The attempt made
slow progress at first.   When sugar was produced from
beetroot, it was not of good quality, nor the price
satisfactory.   Large premiums offered by the Emperor
Napoleon induced the chemists and scientific men of
France to exert their skill; an Imperial factory was
erected ; skill and intelligence were brought to bear on
the subject, and the result was complete success.   It
is now a thriving and increasing industry, not only in
France, but in almost all parts of Europe.   Within
the last thirty years hundreds of factories for producing sugar from beetroot have been erected in France,
Belgium, Austria, Prussia, and Russia ; and thousands
of tons are manufactured annually to meet the local
demand in those countries.   What is more, France is
sending to England sugar made from beetroot, which
comes into competition with cane-sugar sent from India,
Mauritius, and other colonies.   In 1866, in France,
there were 439 beetroot factories, which produced
82,556 tons ; and in Russia there were 424 factories,
which produced 26,519 tons of raw sugar.   Such has
been the result of the progress of improvement in Europe.

In the mother country of the sugar-cane not one step
forward has been taken in its manufacture for centuries.

To squeeze the juice from the canes in several parts of India, two small wooden rollers close to each other are employed. Another form of sugar-mill is on the

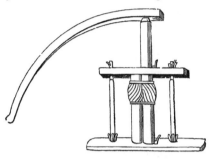

principle of a mortar and pestle. The pestle is rubbed against the canes (cut into thin slices beforehand)—a

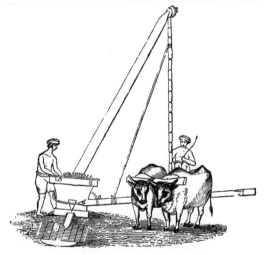

troublesome operation. The moving force is two oxen. The pressure is so imperfect, that a large amount of juice is left, thus causing a loss at the very outset. The juice is boiled in pans heated over an open fire, and is made into *goor* by the poor cultivators. This goor is purchased by persons whose business it is to remove impurities from it and produce sugars of various qualities, known by the names of Khur, Doloo, Gurpatta, and Dobarah, and exported in large quantities to great Britain, and other countries.

Khur Sugar is made in Bengal by pouring goor into coarse gunny bags, and pressing between bamboos lashed together, until 30 or 40 per cent. of it is forced out in the shape of molasses, or sugar that will not crystallize. The residue is Khur. *Ninsphool* or fine khur is made by repeating the above process twice, which causes a further portion to be separated. In both sugars a certain portion of water remains, which renders it liable to sweat. *Doloo* or *Dullooah* is made by pouring goor into open baskets, holding two or three maunds each; three inches thick of wet grass being placed over the goor, the molasses drains through a hole into a vessel placed underneath. As soon as the grass is dry, the upper part of the goor, deprived of its molasses by draining, is scraped off with a knife to a depth of two or three inches, and fresh grass is applied. When dry, a further portion of sugar is scraped off; and this process is repeated till the basket of goor is emptied. The scraped off sugar is placed on mats in the sun to dry. When well made, Doloo is dry, light, and sand-coloured. From a given quantity of goor, 30 to 40 per cent. of this kind of sugar is produced. The resulting molasses, with a small portion of sugar, are melted and boiled, and an inferior dark-coloured goor is produced. Doloo keeps well in dry weather, but not in wet.

Puckha Cheena or *Gurpatta* is the refined sugar of India. It is made by boiling khur with potash temper, which removes the impurities. After skimming, it is filtered through a cotton cloth and boiled, then poured into earthen pots, and as it cools it forms crystals of white sugar. The syrup which drains from the pots is boiled with fresh goor, and an inferior sugar is produced called *Jerannee*. Gurpatta sugar is bright, clean, and dry, and keeps well. The yield from three maunds of good goor is twenty seers of Gurpatta, ten seers of inferior Gurpatta, ten seers Jerannee, one maund twenty-eight seers molasses, and loss twelve seers. *Dobarah* is of superior quality to Gurpatta, being good white, dry, and well crystallised sugar, and is made from Doloo instead of Khur. It resembles the crushed refined sugar of European manufacture.

The juice when expressed from the canes is opaque, frothy, and of a yellowish-green colour, but darkens rapidly on exposure to the air; the amount of acidity increases the longer it is kept, and sugar is then lost more or less. When boiled, a green scum, consisting of impurities, rises, and the liquor becomes of a pale yellow colour. But it has been found by chemists that slow heat destroys sugar, and heat more than 140° F. produces an enormous injury in composition and colour. The Indian system of making sugar is, therefore, very defective. It produces a small quantity of sugar, and abundance of impure goor; hence the cheapness of the rude sugar mills and the cheap process are but little compensation for the mischief and waste incurred.

The average amount of juice in 100 lbs. of cane is about 60 lbs. But the juice consists of a large percentage of water and other impurities; only eighteen or twenty per cent. is sugar, and in practice no more than eight per cent. is extracted in the form of crystallised sugar.

While the process of manufacturing sugar in India, China, and other parts of Asia, has been quite stationary for hundreds of years, in Europe, from its first introduction, improvements have been made from time to time in the mechanical and also the chemical processes. Inventions have been introduced for sugar mills, consisting of heavy iron rollers driven by steam, and extracting the greater amount of juice; for heating by high-pressure steam, with the advantage of not burning any material submitted to its action; for a new mode of communicating heat; for expelling molasses or syrup from the sugar; for producing a high degree of purity and whiteness by charcoal; for a mode of separating molasses and other soluble matters by mechanical pressure; for passing heat through a coil of copper pipe

fixed within the pan containing the sugar solution; for evaporation and concentration of the syrup. An important improvement was patented by a medical man for removing all the impurities in the cane-juice at once,

by salts of lead, and by passing a current of sulphur-
ous acid. In the first year of his patent he sold a
fortieth part of it for £2,000. A London refiner of
sugar, moreover, paid him £1 per ton royalty on his
produce, about £150 per week ; and thousands of tons
of sugar were manufactured by this process. But there
was danger of leaving a poison in sugar if this process was not done well, though it is perfectly capable
of being done well. For this reason it has been dis-
couraged and partly prohibited. Another improvement was that of the centrifugal machine, revolving
about two thousand times per minute by steam power,
in which the crude sugar is placed. A few minutes are
sufficient to separate the white crystalline grains of
sugar, which are retained inside, while the molasses is
thrown out. This machine is now extensively used in
some of the sugar-growing countries, and in England
in the sugar refineries.

From repeated experiments it has been found that
cane juice, if subjected to rapid heat to evaporate the
water, does not lose its colour, nor its property of
crystallising. By slow boiling of the cane juice the first
change effected is the production of grape sugar, which
cannot be crystallised. By further boiling, the grape
sugar is decomposed, and a certain black material is
produced. The injurious effect is in proportion to the
duration of the heat. The mischief does not end here.
The black material produced by continued slow heat
retards again from crystallisation its own weight of
pure sugar contained in the juice : and the loss of sugar
is thus very great. It is to obviate this waste of more
than half the sugar that a machine has been recently
patented in England, and is working (as far as it

has been tried) with satisfactory results in the West Indies, in a sugar plantation. This machine claims the merit of not spoiling the chemical nature of the sugar, or its colour, or injuring the material in any way. Moreover, it is simple in construction, and economical in fuel. The trash of canes after the juice has been extracted in the sugar mill answers for fuel. When the machine is in full action, the cane juice flows in continually at one end, and at the other it is as steadily removed, occupying no more for its passage than fifteen minutes. The product has been called "*concrete*," and the colour of it is a pale greenish yellow. If required for export to Europe for refining purposes, the concrete can be cast in blocks of any size, and packed in bags.

Great Britain, owing to its climate, does not produce a pound of sugar-cane; and notwithstanding the fact that sugar must be manufactured from the cane juice in the countries where it is grown, and that without loss of time, yet there are hundreds of sugar refineries in Great Britain, as well as in all commercial cities of Europe. This had solely its origin in the defective methods employed in the sugar-growing countries, where the fine and white loaf and lump sugar, suited for the European markets, was not made. In England, in sugar refining, occupation is found for a great many establishments for producing loaf sugar, crystal sugar, crushed sugar, and also moist coloured sugars. It is chiefly carried on in London, Bristol, Manchester, Liverpool, Greenock, and Southampton. But Bristol is the head-quarters of this branch of industry, where there are huge sugar factories. The sugar refining works of one firm there, the largest in the world, have cost more than £250,000; in which more than five hundred hands are regularly employed.

A scientific and also very practical English sugar-refiner, who has sugar plantations and sugar factories in the West Indies, has stated that the sugar mill employed in India is so rude that it must leave a large portion of the richest juice in the cane trash; scarcely any machine more miserable can be conceived. The manufacture of sugar also is conducted in the most primitive manner.

It has been said that the climate and soil of India, and other parts of Asia, are all that can be desired for the cultivation of sugar, and that no country holds out a better prospect of success in the prosecution of this important branch of trade, with improved European machinery, coupled with science and industry. In Asia the manufacture of sugar has been unaided by the application of those improved means which have so prospered European enterprise. Asia insists on standing still, while other nations are making progress. The result is that she is poor.

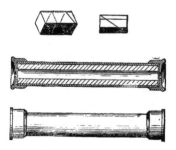

SUGAR FACTORIES.

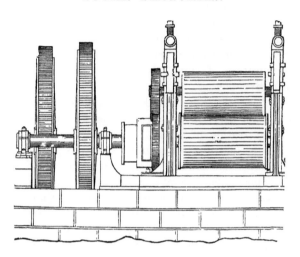

MANUFACTURING PROCESSES.

(1.) *Sugar Mill* has for its object the extraction of the juice from the canes. It consists of large heavy iron rollers, worked by a steam-engine A self-acting cane-carrier is used to supply the canes to the mill, which facilitates the work. The iron feed-rollers split and slightly press the canes, and the other rollers give the final pressure; the juice runs down to the mill-bed, while the trash of canes is carried outside. This trash is used as fuel, and the ashes as manure for the cane plantations. The yield of juice varies according to the completeness of the crushing, and the moisture of the air.

(2.) *Clarifiers.*—The juice expressed from the canes

contains, besides sugar and water, certain impurities, to remove which the juice from the bed of the sugar mill is conducted to large vessels, called *clarifiers*, capable of holding hundreds of gallons each, and heated by steam passing in a coil of copper tubes, regulated at pleasure. The impurities liberated in boiling are conveyed to a cistern, and there are other contrivances for removing

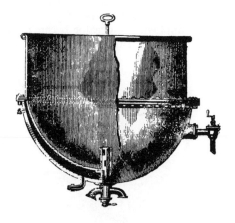

the scum and clearing the juice. Thus the juice is clarified and partly concentrated; and the water being evaporated, it forms a sort of syrup, which passes from the clarifiers into a receiving tank.

(3.) *Concreter* is a machine which has been recently patented for very rapidly condensing the juice. The juice runs zig-zag in a continuous stream into a series of shallow iron trays of this machine, until it reaches the lower end, making a passage in all of 140 feet; and as the trays are heated by a flame from cane trash, it boils rapidly all the time. It is next passed into a copper cylinder 20 feet long, turning six times a minute, and heated outside by waste heat. A strong blast of highly heated air is sent by a self-acting fan through the inside of the cylinder, covering a superficial area of 192 square feet; and by the drying action of the heated air as the juice passes through, the water in a few minutes is driven off, whereby the juice increases in density. The syrup is not at all injured or burnt, and retains its power of forming crystals of sugar.

(4.) *Vacuum Pan* is employed to concentrate the syrup to the point required for forming grains or crystals of sugar. The more rapidly this is done, and the less heat applied, the greater is the production of sugar. A solution of sugar will require 230° F. to

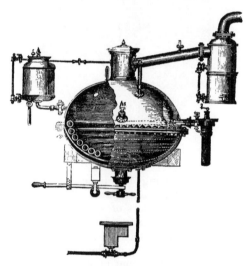

boil; but if the air (which has a pressure) be removed from the pan, it will boil at only 150° F. For this purpose an air-pump is attached to the vacuum pan, and also a coil of copper steam-pipes for heating. The vapour from the boiling solution is condensed by injecting cold water. There is a graduated measure to ascertain how much syrup runs into the pan, and an *air-tight* stick to take out the thick syrup to see when the operation is complete, and which varies according to the size of grain required in the sugar.

(5.) *Centrifugal Machines* are for cleaning and drying sugar. The sugar, with the adhering syrup, from the vacuum pan, is thrown into the cavity of one of these small machines, and is rotated by an engine 1,000 times in a minute. The crystals of sugar are retained in the perforated copper casing of the machine, while the syrup-molasses is forced into the outer case. A rotation of only a few minutes is sufficient for the drying of a charge of sugar, and separating the crystals. The machine is then stopped, and the sugar scraped out. From the syrup or molasses an inferior sugar is made—fine, seconds, or treacle. By means of these machines much time is saved in the draining of sugar ; they are extensively employed in England, in the sugar-growing countries, and in other useful manufactures.

English Sugar Refinery is six stories high. Each story is separated into two divisions—one for melting and boiling, the other for draining and drying.

(1.) Baskets or hogsheads of impure colonial sugar are lifted by a hoist to the top story, dissolved in a pan with water, and stirred mechanically.

(2.) The dissolved liquor falls below from the pans into the filter bags enclosed in a case, and the insoluble matter in the sugar is removed.

(3.) From the bag filters the liquor runs down into a cistern, and the colour of it is of sherry wine.

(4.) The colour is removed and the liquor purified by passing it into a large iron vessel, 20 to 30 feet high, 5 feet in diameter, containing animal charcoal.

(5 and 6.) The liquor, bleached by charcoal, now runs into a tank on the basement floor, from which it is drawn into a vacuum pan to be formed into grains of sugar ; from the pan it falls below into the heater :

(7.) Where the grains of sugar are hardened, which completes its formation. After which it is taken to the centrifugal machines, if the sugar is intended to be in a crystallised form ; if required for pure white loaf sugar, it is poured into iron moulds, and then dried.

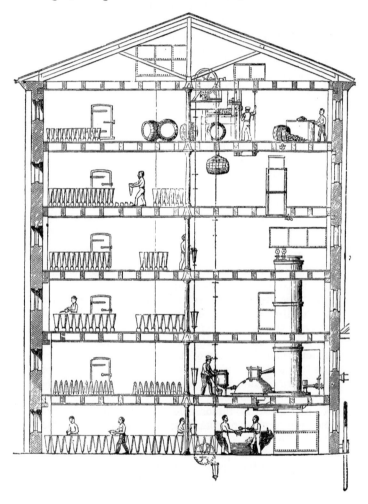

ESTIMATE FOR A SUGAR FACTORY FOR MANUFACTURING SUGAR IN CRYSTALS.

Machinery as shown on PLAN K.

£

A sugar mill, very strong and powerful, with 3 rollers 26″ diameter, 4′ 6″ long, with provision for heating by steam. —Horizontal steam-engine for driving the same, cylinder 15″ diameter, complete, with double gearing; a cane-carrier, megass-elevator, &c., &c. 1,400

3 Steam clarifiers, of 200 gallons each, with internal and external copper tubes for heating, valves, chains, pulleys, weights, gutters for scum, &c., &c. 286

1 Concretor, full-sized, including trays, copper revolving cylinder, drum, self-acting fan, pump, engine, &c., complete, with spare articles, &c. 1,000

Copper vacuum pan, 6′ diameter, 5′ 6″ deep, with jacket, inside dome, and bottom, made of solid copper without seams, with patent seamless copper pipes, condenser, including fittings, with spare proof stick, barometer, &c., &c.—A vacuum pump steam-engine, cylinder 10″, air-pump 16″, feed and cold water pumps, complete.—Cast-iron staging for vacuum pan, with iron stair, rails, plates for flooring, complete (cost, without staging, about £200 less) . . 852

3 Centrifugal machines, for making sugar in crystals, with a separate steam-engine for driving each machine, complete with all fittings 408

1 Iron tank for liquor from clarifier, 1 cistern to receive syrup from concretor, 1 molasses receiver from centrifugals, complete with bolts, nuts, &c. 85

2 Steam boilers, 21′ × 5′ 6″, with 2 flues, complete, with safety-valves, dampers, &c., &c. 475

Steam, feed, and water pipes required in the buildings, including 70 feet for cane-engine, 50 for vacuum pan, &c., 40 for exhaust, 65 for pump of pan-engine, 120 for cane-engine to boiler, 300 for suction and delivery to connect cold-water pumps, 42 for copper canal from clarifiers, &c., &c., including packing and free delivery on board a vessel 705

————

£5,211

M

DIMENSIONS OF A SUGAR FACTORY.

Machinery as shown on PLAN J.

Scale, 20 feet = 1 inch.

	Length, feet.	Breadth, feet.	Square feet.	Height, feet.	Cubic feet.
Room, containing sugar mill, concretor, and other machinery . . .	122	41	5,002	22	110,044
Boiler-house, offices, &c., and store-room	42	30	1,260	12	15,120
Total of square feet and cubic feet			6,262		125,164

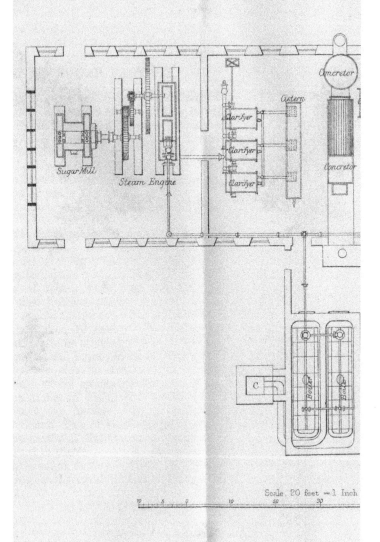

Scale, 20 feet = 1 Inch

J.

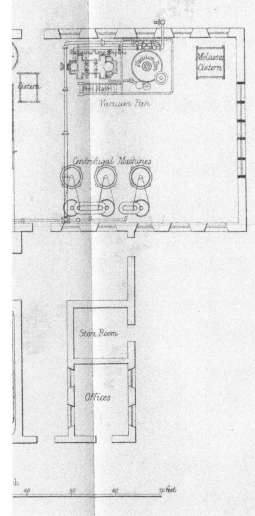

REMARKS ON SUGAR FACTORY.

THE size of a sugar factory will depend on the size of the estate, the number of acres in canes or likely to be, and the amount of work proposed to be done during the season. A sugar factory only works for a certain period during the year; but then it works both day and night. The plans and estimates given are for a moderate sized sugar factory, which will be applicable in a great many cases. The weight of canes from one acre varies, according to soil, culture, and climate, from twenty to thirty tons.

The yield of solid sugar per acre, as a fair specimen, is calculated at 4,000 lbs. But owing to defective culture and the injurious mode of expressing and evaporating the juice in India, the yield in Bengal per acre is only 2,600 lbs. of Khaur sugar, and 2,000 lbs. of molasses, and of such inferior quality, that if Khaur sells per Bengal maund at the rate of $3\frac{1}{2}$ rs., molasses would not fetch more than 10 annas. The means by which a better result is to be attained will be by the use of improved cane mills, and improved machinery driven by steam, which has been done to some extent by enterprising Europeans. Good specimens of sugar made with improved machinery in India in the factories of the Hon. E. Horsman, the Astragan Sugar Company, and Messrs. Caren and Co., of Shahjehanpore, were shown in the International Exhibition of 1862, and medals were awarded to them.

The estimate for a sugar factory includes all machinery for producing well-formed and dry crystals of sugar, and sugar of a quality required for local consumption in India and other parts of Asia, and also for export to Europe. In some parts of Bengal, Europeans have erected improved machinery for manufacturing sugar; but some of the machinery included in the estimate is so recently invented and patented that it has not yet been introduced into any part of Asia.

The plan of the sugar factory is so arranged that machinery for producing pure white loaf sugar may be added whenever required, without any inconvenience. The level, and the connections of the machines with each other, are so adjusted that the juice or syrup, after being expressed, may run down, so as to dispense, as far as practicable, with the use of pumps.

The sugar mill shown on Plan K, the speed of which is so regulated as to produce the greatest amount of juice from the canes, will crush about six tons per twelve working hours. By working it at the required speed, 70 and even 80 per cent. of juice will be obtained.

The concretor will turn out at the rate of 950 lbs. per hour. It has been stated that 1 gallon of cane juice will produce about 2 lbs. of concrete sugar in weight. The total cost of all machinery, buildings, iron roof, charges for erection, and putting in complete working order, may be put down at £10,000. Copper enters largely into the composition of sugar machinery, and therefore the machinery is costly.

Rum is distilled in some sugar factories from molasses and other impurities; but large copper stills and refrigerators are required for the purpose; and with improved machinery for manufacturing sugar, in a great many cases it will not be worth while to make rum.

It is a fact that sugar refineries in England not only produce pure white loaf sugar, but also an equal quantity, perhaps more, of crystalline sugar of every quality sold in the shops in England, as it pays better. This is produced from sugar (imported from India and other sugar-growing colonies) which is more deeply coloured. Disinterested persons have expressed in the English journals their decided opinion that it would be more economical and rational to produce sugar of the requisite tint in the place where sugar-cane is grown and sugar is manufactured, at once, and then export it; than that it should undergo two processes, almost similar, in the colonies and in England, as is the case at present for producing crystallised sugar.

For pure white loaf sugar, additional machinery will be required beyond that specified in the estimate, consisting of bag and charcoal filters, &c., which will cost about £1,300 additional. In a factory where the operations of sugar refining are to be pushed to the highest point for producing pure white loaf sugar, a very large quantity of animal charcoal is quite indispensable. Charcoal entirely removes colour from the sugar solution, making it as colourless as pure water. In England one ton of charcoal is used to purify one or two tons of sugar, according to quality. But, unfortunately, in a very short time, from 24 to 72 hours, the power of the charcoal becomes exhausted. It is restored again by heating it to redness in expensive furnaces. Thus, not only is a large quantity of charcoal required for producing pure white loaf sugar, but it is required to be repeatedly burnt. A sugar refinery for producing pure white loaf sugar will not pay in India unless fuel should become very abundant and cheap.

As a rule, the pure white loaf or lump sugars cannot with advantage be turned out in sugar-growing countries; but any varieties short of those qualities could, under a proper application of machinery, be more profitably manufactured in India, and other parts of Asia, than in England.

IMPORTS OF RAW SUGAR FROM INDIA AND THE EAST INTO GREAT BRITAIN.*

From	1866.	1865.	1864.	1863.	1862.	1861.	1860.	1859.	1858.	1857.	1856.
	cwts.	cwts.	cwts.	cwts.	cwts.	cwts.	cwts.	cwts.	cwts.	cwts.	cwts.
Bengal	360,491	498,029	481,350	14,315	101,006	419,637	364,976	550,126	463,659	661,527	655,413
Madras	"	"	228,223	218,566	207,710	250,228	289,034	283,080	255,421	422,175	460,355
Bombay	"	"	1,553	343	45	1,163	1,038	868	626	9,724	26,353
Penang			Included with Singapore,				76,044	61,034	53,117	45,068	69,848
Singapore			96,637	60,051	98,641	79,203	29,779	34,119	18,484	41,702	15,876
Mauritius	993,680	980,296	1,064,429	1,635,671	686,433	1,503,961	1,163,732	1,169,341	1,086,501	1,184,329	1,647,257

EXPORT AND VALUE OF SUGAR AND SUGAR CANDY FROM INDIA TO GREAT BRITAIN.†

Year.	Quantity.	Value.	Year.	Quantity.	Value.	Year.	Quantity.	Value.
	cwts.	£		cwts.	£		cwts.	£
1865		765,110	1859	1,134,876	1,450,767	1854	1,094,821	941,582
1864	283,568	716,857	1858	962,128	1,175,771	1853	1,477,673	1,729,763
1863	708,687	312,042	1857	1,568,571	1,786,077	1852	1,607,508	1,801,660
1862	845,871	826,936	1856	1,277,060	1,359,104	1851	1,591,631	1,823,789
1861		1,032,416	1855	958,563	1,135,699	1850	1,624,376	1,925,602
1860	860,001	1,031,944						

* From Reed's Sugar and Sugar-yielding Plants Returns for 1865-66, from Board of Trade Returns.
† From Statistical Abstracts presented to Parliament.

MANUFACTURE OF OILS.

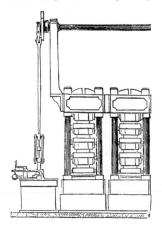

INTRODUCTION.

OIL seeds are valuable and interesting substances,
used for a variety of purposes. The trade in linseed,
rapeseed, and other seeds, and in cocoa-nuts, between
India, and England and France, has during the last few
years increased enormously, especially since the Crimean
war. The oils most extensively used in Europe are
linseed, rapeseed, cocoa-nut, palm, and castor oils.

The vegetable oils are for the most part obtained
by expressing, the seeds being previously ground or
bruised. Some seeds after this treatment readily yield
to pressure, and give what is known as cold-drawn oil;
others give up their oil with difficulty, and it becomes
necessary to heat the bruised seeds previous to pres-

sure; then the oil becomes more easy of separation. The means adopted for expressing oil from seeds and nuts up to the present time, as may be seen on the Malabar Coast and in other parts of Asia, are the same as those adopted in the early stages of civilisation in Europe. The apparatus is very primitive, consisting simply of a few poles driven into the ground and supporting two cross bars, between which a bag containing the seed is placed. A lever is brought to bear, causing pressure upon the seed, from which the oil is expressed. This rude apparatus is slow and inefficient. In England, in the first instance, stampers were introduced from Holland; then the screw-press, and lastly the hydraulic press—now in general use. With stampers it was necessary to work the seeds twice over, whereas with the hydraulic press it is only necessary to work them once. The saving of labour is 25 per cent., and the yield of oil is the same. For all practical purposes it has been found, from actual trial, that the screw-press and the stamper are far inferior to the hydraulic press. But even in the latter, with the emptying and discharging, the process is constant and laborious. Oil machinery, even the best now in use in Great Britain, is not so perfect and self-acting as machinery in other branches of industry; but still Asia ought to take every advantage of the improved machinery now in use in the most civilised parts of the world at the present time, instead of the rude apparatus at present employed for the purpose of expressing oil.

When oil is expressed with rude and imperfect machinery, it becomes foul, rancid, and discoloured in a long sea voyage, and its value diminishes; but if carefully pressed with improved machinery it remains fresh, and fit for the purposes in the arts and manufactures to which it is applied.

Linseed Oil is obtained by expression from the seeds of the common flax plant. In the Bombay Presidency, and in other parts of India, the plant is largely cultivated exclusively for the seeds, and not for the flax-fibres from which linen and other fabrics are made. For the production of linseed a hot climate is very favourable, as the yield is greater, and the seeds larger, plumper, and more oily than in temperate countries. Seed-crushers in England find the flax-seed of India the most productive, as it contains six per cent. more oil than the seeds imported into Great Britain from Russia. Cold-drawn linseed oil of a light yellow colour is better than that expressed with heat; but by heating, the seeds yield seven per cent. more oil. Linseed oil is the most useful and important of all the drying oils. It attracts oxygen from the air and solidifies, and this property renders it valuable for the purpose of making varnishes and paints. It is used also in the manufacture of printers' ink, by allowing it to burn for some time, and then mixing it with lamp-black. The seed exported to England from India is all consumed by the oil mills. After expressing the oil, the seed-cake which remains is most extensively employed for feeding cattle. Linseed cake forms an item in the imports from America and from certain European countries into Great Britain.

Rape Seed Oil, after being purified, is largely used in England for lubricating machinery; for which purpose it is in great demand, the demand increasing with the extension of machinery. For stationary or locomotive engines where high speeds are required, it is stated to be particularly adapted. For marine engines, these qualities, combined with absence of odour, render it much valued. Some idea of the consumption may be formed from the fact that one single Railway Company in England, the London and North-Western, consumes annually more than 40,000 gallons of this oil, for lubricating their locomotives alone, which quantity it is stated would be equal to the production of 1,000 acres of land.

The *Castor oil* plant is also a native of India, where it grows to a considerable size, and lives for several years. The oil is obtained from the seeds by expressing. In Calcutta it is prepared by crushing it between rollers, then placing it in hempen cloths, and pressing in an ordinary screw-press. The oil is then heated in a boiler with water, strained through flannel, and put into cans. The best East India castor oil is sold in London, as "cold drawn."

Cotton Seed Oil is now an important article. There are several mills in England for producing it, and cotton seed-cake as food for cattle. The oil is used in the manufacture of soap. Owing to the enormous quantity of cotton produced in India, there is always an abundance of seed, which it would be desirable to turn to more profitable purposes than is done at present. In America the seeds are pressed near the cotton plantations, as they are bulky, and the cost of transport is necessarily great. The husk and attached fibre are used in paper-making, and the cake after expressing the oil is nearly as valuable as linseed cake.

Cocoa-nut Oil, expressed from the fruit of the palm-trees growing so luxuriantly in the south of India and other parts of Asia, is principally employed in Europe for the manufacture of candles and soap. It is bleached by acids, or by the combined influence of air, heat, and light. A white crystalline fat is separated, for making candles; also a glycerine, which is extensively used now-a-days in the arts, and even in medicine; and the residuum of the oil is converted into a fine hard pitch, fit for any purposes to which ordinary pitch is applied. The soap made from cocoa-nut oil is very white and light, and to a larger extent soluble in water than any other soaps; it is useful in washing, also, with salt-water.

The *Tallow Tree* has been introduced from China into India lately, and flourishes with great luxuriance in the Dehra Dhoon, in the Dhoons and Kohistan of the North-Western Provinces, and in the Punjaub; and there are now tens of thousands of such trees in the Government plantations. The tallow prepared from the seeds has been tested on the Punjaub Railway, as a lubricator for locomotive machinery. The leaves of this plant also make valuable dye.

A large quantity of oil is used in the manufacture of soap and candles in England, and other countries. Price's Candle Company possesses cocoa-nut plantations in Ceylon, and employs eight hundred workmen in its several manufactories in London, using a capital of nearly half a million sterling, and dividing profits to the extent of £40,000 per annum. Till quite recently, the linseed oil required for Government use, throughout India, was wholly sent out from Europe; and it is only within the last few years that it has been found out that the native grown linseed is quite as good as the best which can be had in Europe. In a

country where the people are so extremely fond of
retaining old methods, and so slow to recognise the
science and improved methods of European countries,
it would be better if Government were to give more
stimulus and inducement to new branches of industry.

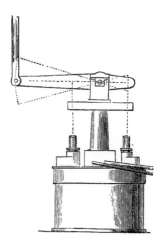

OILS.

MANUFACTURING PROCESSES.

(1.) *Seed Crushing Rollers.*—A self-acting elevator, as it revolves, takes a certain quantity of seeds each time, by small buckets, from the heap to a flat screen or shaker, which is kept constantly moving to clear the seeds of all foreign matter. The seeds are caused to descend equably between two heavy crushing iron rollers, revolving in opposite directions, and so adjusted as to take in different sizes of seeds. A pair of rollers

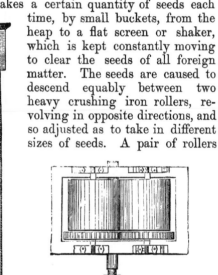

will crush on an average 4 tons of seeds per day. For cocoa-nuts a rasping machine is employed, for cutting and reducing the *copra* to such a consistency as to prepare for the second process.

(2.) *Grinding the Seeds.*—The

second operation consists in grinding the seeds under a
pair of edge stones, weighing about 7 tons. In the
circular path of the edge stones the seeds fall from the
rollers; and as they go round about seventeen times
per minute, the seeds are crushed not merely by the
weight of the stones, but also by a rubbing motion.
In about half-an-hour the seeds are sufficiently ground
into a paste. A knife revolves with the stones for turning
up the mass and detaching that which adheres. The
stones, if of good quality, will last for a great length
of time, but will require to be faced every three years.

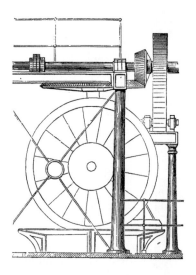

One pair will grind sufficient seed for two double
presses. The stone generally used is the best Derby-
shire, furnished with iron plate and ledge many inches
deep.

(3.) *Heating Seeds.*—The pasty mass of crushed seeds

is generally heated in a double steam kettle, which consists of two round chambers, one over the other. Round the sides and bottom is the space for admitting steam; and in the middle of the pan are two arms or stirrers, revolving 36 times a minute, which keep the seeds agitated, so that every particle may come in contact with the heated sides. After a few minutes, when the bruised seed is sufficiently heated, the double door is opened, and the contents discharged by the action of the revolving stirrers. The heating operation is continuous—by first charging the upper chamber, then the lower, where the seeds remain till required. The seeds fall by an orifice into a bag.

(4.) *Bags*, for receiving the heated paste of the seeds, and placing in the hydraulic press, are made of strong cloth woven on purpose. After being filled, they are placed separately between what are called the *hairs*, which are bags made of horse hair, with an external covering of leather. The bags are expensive, and on account of the great wear and tear are required to be renewed every year. One set of sixteen will cost about £135.

(5.) *Expressing Oil.*—For expressing the heated seeds the hydraulic presses used are *double*. Each single press is fitted with four boxes, and receives four bags of heated seed paste. The attendant first fills one single press, opens the communication, and brings a pressure to bear; and while this is going on, the second single press is being filled, the valves of the press pumps are opened, and a total pressure of about 300 tons is exerted. The oil from the seed paste passes through the canvas bag, then through the hair bag, then round the upper portion of each box, and by pipes goes into a cistern. The oil cakes are put into a kind of rack to cool and dry, so that they may not become mouldy when stored. For emptying and charging the press, and for expressing the oil, it takes about ten minutes.

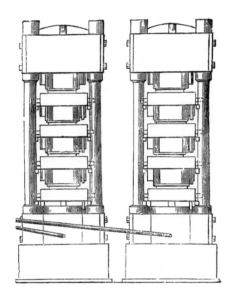

ESTIMATE FOR AN OIL MILL.

Machinery as shown on PLAN I.

	£
Elevator for elevating oil-seeds, complete; a pair of rolls for crushing seeds, diameter 27″ and 21″, complete . . .	173
Cocoa-nut rasping machine, with knives and fittings complete (only required for cocoa-nuts, not for seeds) . . .	96
Grinding-stones, for grinding seeds, best Derbyshire, furnished with plate and ledge 7″ deep, fitted with shaft diameter, weight 10 tons, complete	138
Steam-kettle for heating seed-paste, furnished with stirrers, 100 feet steam-pipes, pressure-gauge, &c.	52
Bagging for containing heated seed-paste for pressing, 32 hairs with double leather backs, including one extra set, complete	264
Two double hydraulic presses for expressing oil, with 8 boxes in each double press, or 16 boxes in all, diameter of rams 12″, including a set of patent double-pumps, with indicator, complete. (If required extra strong, weighing 26 tons, add about £165.)	527
Four oil cisterns 6′ × 8′ × 8′, with glass indicators, a force-pump for forcing oil from the presses to the stock-cistern, 68 feet pipes, with small cistern, complete	192
High-pressure steam-engine, cylinder 16″ diameter, stroke 2′ 6″, with expansion-valve, governor, force-pump, water-heater, &c.; boiler, furnished with patent tubes, complete, with water-gauge, safety-valves, and all fittings	520
Gearing, including shafting, 6½″ diameter, spur and bevel wheels, pulleys, independent iron framing, iron columns, heavy iron girders, complete	360
Miscellaneous and extras, including cake-tables and strippers, leather straps for hairs, leather rings, copper rivets, &c. (Packing and free delivery on board)	200
	£2,522

N

DIMENSIONS OF AN OIL MILL.

Machinery as shown on PLAN K.

Scale, 10 feet 8 inches = 1 inch.

	Length, feet.	Breadth. feet.	Square feet.	Height, feet.	Cubic feet.
Room containing machinery for expressing oil and engine	46	26	1,196	14	16,744
Boiler-house	22	17	374	10	3,740
Seed store-room	20	14	280	13	3,640
Offices	20	14	280	13	3,640
Cake warehouse*	14	26	364	12	4,368
Oil-tanks room*	18	26	468	12	5,616
Total of square feet and cubic feet			2,962		37,748

* Not shown on plan.

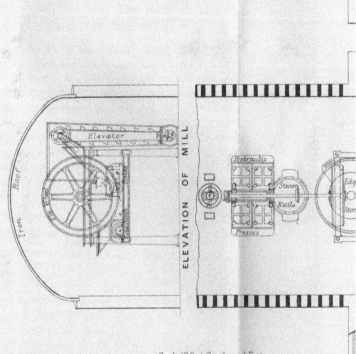

ELEVATION OF MILL

Iron Roof

Elevator

Fly Wheel

Hydraulic

Steam

Kettle

Presses

Edg

Ston

Scale 10 feet, 8 inches = 1 Foot

0   5   10   20   30 feet

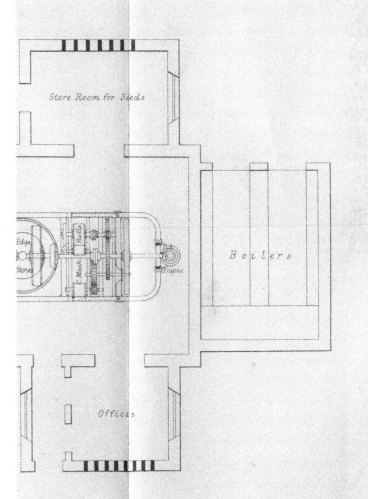

Store Room for Seeds

Edge
Stones

Rolls

C.Mach.

Engine

Boilers

Offices

REMARKS ON OIL MILL.

THE oil mill for which the estimate and plan are given is on what is called the self-supporting principle; that is, all the machinery is self-supporting on independent iron framing, consisting of columns and heavy girders, and worked quite independently of any buildings. The machinery consists of two double hydraulic presses, each double press with eight boxes, or sixteen boxes in all, with preparatory machinery adapted for crushing every description of oil seeds, as well as cocoa-nuts.

The *yield of oil* from the seeds will vary according to the quality of seeds.

1 Quarter of	Bombay linseed yields from	125 to 140 lbs. of oil.
1 ,,	Calcutta ,,	,,	118 to 130	,,
1 ,,	Russian ,,	,,	108 to 116	,,
1 ,,	Bombay rapeseed	,,	120 to 145	,,
1 ,,	Guzerat ,,	,,	140 to 180	,,
1 ,,	Calcutta ,,	,,	115 to 135	,,
Cocoa-nut		,,	60 to 65 per cent.
Cotton seed		,,	16 to 20	,,

Linseed is worked once over, rapeseed twice over. Calcutta and Bombay seeds require to be watered in working, whereby they gain from 3 to 10 lbs. What remains after pressing the oil is called the oil-cake.

Oil-cake.—There is always so much seed paste put into the boxes of each hydraulic press, that, after the oil is expressed from it, each cake will weigh 8 lbs. In an English oil mill one man turns out a set of cakes from each double press in ten minutes, consequently each double press with eight boxes makes 80 lbs of cake in ten minutes. The use of oil-cake for feeding

cattle increases every year in Europe; it is an article
of import into Great Britain from America and other
places, and a large quantity is also manufactured in
England.

Plan K shows the arrangement of the oil machinery.
The store-room for linseed is of a size to hold 200
quarters, about a week's stock; but the size may be
doubled.   The oil-tanks and oil-cake house are not
drawn to scale on plan.   The oil is pumped from the
presses to the iron tanks; the stock tanks may be
placed at any distance, and in any position as regards
the mill, and connected by pipes.

Wages.—In an oil mill in England wages range
from 24 to 10 shillings per week.   The foreman is paid
30s. per week.   No more than ten hands are required
in a mill of two double presses.

As per estimate, the machinery with two double
presses will cost £2,500, and, including freight, insurance, cost of buildings, iron roof, erection of machinery, &c., the total cost of the oil-works may be put
down at less than £6,000.

Production.—The oil mill, with two presses and
sixteen boxes, for which the estimate is given, will, by
working ten hours a day, crush 9,360 quarters of
Bombay linseed per annum.   The production of linseed
oil will be say 564 tons, and of linseed cake 1,174 tons.
A ton of oil, linseed or rapeseed, contains 240 gallons,
and occupies 38½ cubic feet.   A cistern 8 × 8 × 6 feet
will contain 2,393 gallons.   Oil mills in England generally work day and night.

Profits.—The following table shows the amount of
profit derived, after deducting all expenses, wear and
tear, by one of the oil mill proprietors in England, in

five successive years, by working Bombay linseed in two double presses, sixteen boxes in all :—

Linseed bought per quarter.	Oil sold per ton.		Oil cake sold per ton.		Profits 12 months.
s.	£	s.	£	s.	£
52	28	5	9	10	1,528
56	30	0	10	0	1,230
72	45	0	10	0	2,202
66	37	10	9	10	1,193
59	31	0	10	10	977

As an example of the fluctuations in profits, and the difference in pressing different seeds *in the same year*, the following figures will suffice, the amount of machinery being the same :—

Cotton seed bought at .   .   .   . £8   0s. per ton.
  ,,   oil sold at   .   .   . £31 0s.   ,,
  ,,   cake sold at   .   .   . £5   5s.   ,,
        Profits realised in one year £4,046.

Rape seed bought at   .   .   . £3   7s. per quarter.
  ,, oil sold at .   .   .   . £49 0s. per ton.
  ,, cake sold at   .   .   . £5 15s.   ,,
        Profit realised in one year £2,333.

The *Bombay Oil Mill* was started for pressing cocoanuts, but it failed through special and easily explicable causes. A European was engaged in the capacity of mill-manager, and sent over to England, and a respectable firm was instructed to purchase machinery according to his recommendations, which was accordingly done. The oil mill consisted of two double presses, and other machinery for pressing oil, the same as specified in the estimate, made to fit a particular building at Goa. A timber mill was also attached to it, to be worked by one engine. When the machinery arrived it was erected, not at Goa but in Bombay; the timber mill

was left out; and before the oil machinery commenced
to work, the European manager had already received
for his double passage to and from India, and as salary,
more than £1,000, which was a pure waste.   Besides,
the steam-engine power was three or four times more
than actually required.   Under these circumstances
failure was inevitable.

All oils and fats, when treated with soda or potash,
yield soap.   The general method of soap-making is the
same for all kinds, though it varies in some particulars
according to the kind of oil or fat.   Soap is manufactured in cast-iron pans built in brickwork, into
which steam pipes are introduced for heating the oils
with potash or soda.

Saint Brothers have an establishment at Cossipore,
where stearine candles are made, which are hard, white,
and dry, and were shown at the International Exhibition in 1862.

OIL SEEDS AND OIL EXPORTED FROM INDIA TO GREAT BRITAIN.*

Year.	Seeds, qrs.	Value—Seeds, £	Value—Oils, £
1865	—	1,912,433	217,730
1864	—	2,032,832	422,175
1863	844,090	1,833,851	372,107
1862	582,768	1,206,331	209,502
1861	983,882	1,785,526	247,094
1860	1,129,167	1,548,721	180,066
1859	—	2,059,445	192,562
1858	—	1,380,001	265,271
1857	364,506	1,118,654	179,164
1856	—	1,273,457	154,540
1855	611,655	812,799	130,958
1854	311,936	471,797	104,170
1853	282,480	448,770	90,039
1852	309,982	501,420	92,722
1851	218,762	339,514	129,121
1850	130,243	216,510	106,947

* From Statistical Abstracts presented to Parliament, 1867.

COCOA-NUT OIL EXPORTED FROM CEYLON TO GREAT BRITAIN.*

Year.	Quantity, cwts.	Value, £
1864	180,761	224,955
1863	152,076	189,232
1862	115,084	143,216
1861	83,608	104,042
1860	129,698	161,403
1859	98,006	121,964
1858	66,001	82,134
1857	142,025	223,254
1856	87,396	105,383
1855	99,168	144,227
1854	98,218	138,547
1853	99,914	116,017
1852	70,848	69,209
1851	38,828	34,371

* From Statistical Abstract of Colonial and other Possessions, 1866.

OIL GAS.

In places where coal is difficult to procure, oil may be advantageously substituted as a source of gas for illuminating purposes. The oil employed for this purpose is the crudest and cheapest that can be procured; even such as is quite unfit for burning in the ordinary manner is sufficiently pure for making oil gas The illuminating power of oil gas is *twice* as much as that of ordinary coal gas. The process of making it also is much more simple, as it requires far less purification, and the production of gas is large compared with coal gas. In Liverpool, Bristol, Hull, and other places in Great Britain, extensive establishments were at one time erected to produce gas from oil;

but as coal is so cheap, and the commonest oil far too expensive, the manufacture of gas from oil has been discontinued. But in several places in India, and other parts of Asia, the manufacture of oil gas will be advantageous, besides being an admirable means of using up impure oil, refuse fat, and such other materials for the production of light as could not otherwise be employed, or could only be applied to the lowest uses.

Oils, fatty substances, and coal, contain a very large portion of compounds and other ingredients required for illuminating purposes. But combined with coal or oils, there are other substances which are not required, which do not contribute to the illuminating power, and which may be separated when gas is manufactured. Coal contains more impure substances than oils; hence the manufacture of gas from coal requires more purification than that from oils; besides which, the principal illuminating ingredients predominate more in oils than in coal, which is one great advantage in manufacturing gas from oils.

The superiority of gas over the dim oil lights was seen in Bombay so late as 1866, and in Calcutta in 1860, when it excited at first as much popular attention as it did in England when the first public display of gas was made in the year 1798. The time seems not far off when every town, even in a country like India, will be lighted with gas, and the dim oil lamps entirely displaced. For brilliancy, nothing practically surpasses the light from gas. Within the last few years gas has been also extensively used for cooking food, and for other heating purposes.

The late Framjee Cowasjee—a gentleman whose name will always be held in grateful memory as one who, without having had the advantage of receiving a good education, did more to introduce new industrial enterprises in Western India than any other native of India has yet done—when he lighted his mansion with gas, and (to celebrate the event) gave a dinner to his friends,

some were made sick by the noxious smell, seeing that
at that time no improvements had been effected in
purifying the gas.   But now gas apparata are made,
not only for large towns, but of any size and for any
number of lights, for placing near factories, gentlemen's
mansions, and other public buildings, without producing any unpleasant odour, and made so simple in
construction that any person could work them with ease.

Notwithstanding the present entire absence of anything like gas in the interior towns in India, and other
places in Asia, there is little doubt that its manufacture
from refuse oils, or other comparatively useless fatty
substances, requires only to be developed to be duly
appreciated.   Once introduced, its consumption will
increase, and in proportion as attention is paid to improvements in its manufacture, in the fittings, and in
the mode of burning, it will become more and more
generally adopted in large towns, and even in private
houses, so as to supersede at last almost all other
sources of artificial light.

In the manufacture of gas, the first process is to heat
the material in a close iron retort by a furnace in the

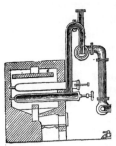

usual way.   The oil is allowed to
flow in a continuous stream from a
tank to the retorts, in which small
pieces of brick are introduced to
increase the heating surface, and
to shorten the time for the production of gas.   The oil is decomposed by a slow red heat, and is
converted into illuminating gas
and tar, which both escape as
vapour by a pipe.   The tar gas,
being heavy, deposits the tar in a small cistern, which
is always kept half full; while the other gas is conducted away to be purified.

The gas which escapes by the pipe from the oil retort contains impurities which require to be separated

from it. Lime is used for purifying, and the gas is passed into a suitable vessel containing it, which takes away the sulphuretted hydrogen, carbonic acid, and other matters. The gas is then stored in a gasholder, ready to be distributed by pipes in the places to be lighted. In the bottom part of the gasholder is a tank of water; in the upper is the vessel of iron suspended by chains, moving up and down freely. The gas, from the force it acquires by being expelled from the oil, presses on the surface of the water, and causes the iron gasholder which holds the gas to rise; and in proportion to the quantity entering, so does the holder rise out of the water.

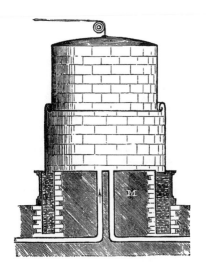

PRICES OF OIL GAS AND DOUBLE RETORT APPARATUS,

COMPLETE,

INCLUDING RETORTS, PURIFIERS, GASHOLDERS, AND FIRE-BRICKS.

No. of Burners.*							Price. £
6	.	.	.	.	.	.	26
10	.	.	.	.	.	.	58
20	.	.	.	.	.	.	75
30	.	.	.	.	.	.	92
40	.	.	.	.	.	.	120
70	.	.	.	.	.	.	195
100	.	.	.	.	.	.	231
200	.	.	.	.	.	.	370
300	.	.	.	.	.	.	480
400	.	.	.	.	.	.	588
500	.	.	.	.	.	.	781
800	.	.	.	.	.	.	1,061
1,200	.	.	.	.	.	.	1,225
1,600	.	.	.	.	.	.	1,573
2,000	.	.	.	.	.	.	1,925

Each burner will consume $1\frac{1}{2}$ cubic foot of oil gas per hour; for the same intensity of light 3 cubic feet of *coal* gas would be required. One gallon of oil will yield 80 cubic feet of rich gas.

The price of gas in London is 4s. per 1,000 cubic feet.

In Calcutta, the price charged by an English company is 12s.

The cost of oil gas in India, as tried for several months, is 16s. 6d. per 1,000 cubic feet from cotton-seed oil.

* Each burner burning on an average three hours.

PAPER MANUFACTURES.

INTRODUCTION.

OF all the useful arts there are few
which contribute so much to human
happiness, and to elevate and improve mankind, as the arts of writing, printing, and paper making. In
the primitive ages people were satisfied with writing on stone, metals,
and bricks.   Subsequently more
pliable textures were employed, such
as skins of sheep or goats, and leaves
of trees. Books formed of such leaves
have been preserved in the libraries
of Europe ; and the practice of writing upon leaves of trees is still in
use in some parts of Asia up to the
present day.

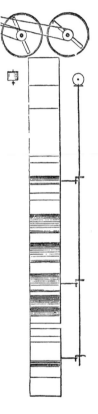

The invention of the paper manufacture is most valuable and useful,
as it saves infinite trouble, labour,
and time; and without it every other
discovery would have continued
useless to society. The civilisation
of mankind depends mainly on the
diffusion of knowledge, and paper is
the most essential medium for dif-
fusing that blessing to the human
race.

China originally gave birth to
the invention of paper.   The art
of making paper from vegetable matter reduced to
pulp was known and understood in China centuries

before it was practised in Europe, and carried to some perfection. There is no doubt that the novel idea of reducing a vegetable to a liquid state for the purpose of obtaining by deposit a smooth thin flexible sheet is due to the Chinese. These ingenious people make paper from cotton, hemp, rice straw, a species of mulberry tree, and from the bark of the bamboo. The young stalks of the bamboo are cut and thrown into a reservoir of mud and water for a fortnight, to soften them, and then taken out, cut, and exposed in the sun to dry and bleach. After this the pieces are boiled in large kettles, and reduced to pulp in mortars by hand. Some glutinous substance is mixed with it, and the whole beaten together to the required thickness with water, and placed in a vat. A workman dips a sort of framed sieve into this vat, and on raising it, and shaking it, the water passes off through the holes of the sieve, and the stuff remains on the surface, whereby a sheet of half-formed paper is made. This is pressed against a smooth surface near a stove, so that the sheet of wet paper may adhere. By the warmth of the stove the water evaporates rapidly; and before the paper is quite dry it is brushed over with a size made from rice, which soon dries, and the paper is then stripped off in a finished state. Such is the rude process by which not only in China, but in India and other parts of Asia, paper is made, with some slight modifications, up to the present time.

In Europe the manufacture of paper was introduced only in the fourteenth century. The first paper mill in England was erected in 1588 by a German, for which he received the honour of knighthood from Queen Elizabeth, and a licence was also granted to him for the sole gathering of rags for making paper for ten years. Till 1765 scarcely any high degree of perfection was reached. Paper made in Europe by hand at that time is thus described—the Venetian as neat, subtle, and court-like—the French light, slight, and tender—

the Dutch thick, corpulent, and gross, sucking up ink.

Up to the end of the eighteenth century paper was made by hand in Europe. The honour of inventing that beautiful contrivance, the *continuous* paper machine, which has contributed so much to the advancement of civilisation, is due to a Frenchman. It was only in 1803 that the *first* paper machine for making paper continuously was worked in England. Chlorine gas was soon adopted, by which the deepest coloured rags for making paper were deprived entirely of their colour, and rendered quite white. In place of the old fashioned mortars and mallets, engines were established. In 1840 sand-traps for catching gravel and other impurities in the pulp were used; and in the same year suction pumps were employed to carry off by degrees a great portion of water contained in the pulp from the wire-cloth on which paper is made. To perfect the continuous paper machine, one single firm laboured for six years, and expended £60,000 on it. The amount was so large that they became bankrupt, as their patent, owing to some technical objections, was declared void. By the improvements effected by the firm the cost of making paper was reduced from sixteen shillings to three shillings and ninepence per hundredweight. In 1849 paper-making advanced still more rapidly owing to improvements in machinery. The composition of pulp was better studied. Between 1852 and 1857 no less than one hundred and forty-seven patents were taken out in England for improvements in paper, but chiefly for new materials for the pulp.

The principal improvement in paper-making by machinery is the adaptation of an endless revolving wire cloth to receive the paper pulp with an endless felt, to which in its progress the paper is transferred by self-acting motion; this is so quickly done that at one end of the paper machine the wire cloth receives

the constant flow of liquid pulp, and in the course of
two or three minutes at the other end of the machine
it comes out in the form of finished paper.

The largest paper that has been made by hand is
53 by 31 inches ; and so great is the weight of pulp
for making a single sheet of this size, that no less than
nine men are required ; whereas with the continuous
paper machine, not one sheet only, but a continuous
sheet of any length, and eight feet wide, is drawn out
and wound on reels. What is more ; the work which
occupied three weeks by the old method by hand is
now done in as many *minutes* by the improved paper
machine. Paper may be made by this machine considerably less than a thousandth part of an inch in
thickness ; and although so thin it is capable of being
coloured and glazed.

The raw material used in England for making paper
is chiefly worn-out rags, which would have little value
if they were not used for that purpose. Cotton and
linen rags are the best, though hemp, jute, and rags of
other fibres are used for mixtures. Cotton and linen
rags afford the greatest facilities for manufacture, and
nearly all are made into white paper. Old ropes, old
cotton bagging, and the sweepings of cotton mills also
furnish a large supply for wrapping papers. Spanish
grass and straw are also used in English paper mills as
an auxiliary to rags, and in conjunction with cotton or
linen rags. It has been proved that paper can be made
from any vegetable fibre which will combine into a pulp.
The only question is the cost of production of paper from
that material. England requires upwards of 120,000
tons of rags for making paper every year, a large proportion of which is imported from other countries.

For several years past a newspaper has been printed
in Sweden upon paper made from horse dung. Some
of the soluble parts in its manufacture are carried off
and used for manure.

In America, at a time when war was raging between North and South, and the supply of cotton rags fell short, great attention was paid to other substances for manufacturing paper, and patents were taken out, and worked successfully, for making paper from wood. The establishment of the American Wood Paper Company, in America, covers ten acres, and their works were completed in 1866 at a cost of 1,000,000 dollars. They are the largest in the world for the manufacture of wood pulp, producing 34,000 lbs. of pulp per day. In other works the wood is first cut into fine chips by machinery, then boiled in a chemical solution in closed boilers, which renders it soft. From the pulp the chemical solution is drained, and conducted by pipes into furnaces, where by a patent process eighty-five per cent. of the substance used in boiling is recovered and used again. It has been stated that without this patent process of recovering the chemical used, the manufacture of paper from wood pulp would be a failure commercially. With the wood pulp a portion is mixed of straw pulp; and paper thus made at these American mills without any rags is of sufficiently good quality to be used for printing newspapers.

In the United Kingdom there are no less than 430 paper mills spread throughout the whole country. Some idea of the extent of these works may be formed from the fact that in one of them, from the very refuse of our clothing and rags, there is made on an average every day forty miles of paper in length and more than five feet wide. Paper is an important branch of trade in the United Kingdom, already worth six millions sterling, and employing a large number of hands.

In the United States, in 1860, there were more than 555 paper mills, producing paper of the value of 17,148,194 dollars. But not only in Great Britain and America; in France, Austria, Denmark, Spain, and other countries in Europe, paper mills turn out thousands of tons of paper every year.

In India, China, and other parts of Asia, paper is made in large quantities, but by *hand*. Over all Asia there are hardly half-a-dozen continuous paper machines working at the present time. Asia has an enormous population, producing vast quantities of cotton clothing, which, though worn more than in Europe, is still of great value for making paper. The supply of raw material is abundant. It behoves her to take every advantage, as all European nations have done, of the recent applications of improved machinery, by which the facilities for producing *cheap* paper have increased so much, and so great a saving of expense, trouble, and material has been effected.

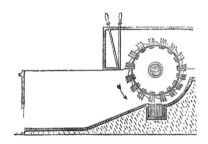

PAPER.

MANUFACTURING PROCESSES.

(1.) *Rag Cutting.*—The rags are cut into small pieces about four inches square, by women, who stand at a table the upper part of which is covered with a netting of coarse wire, a large knife being fixed in the centre with which they cut the rags, and keep different qualities of rag in the several divisions of the box.

(2.) *Rag Dusting Machine.*—The rags are put into this machine, which is a large cylindrical cage or drum, covered with iron wire, six feet in diameter, and having a revolving shaft with a number of spokes in the interior, which toss and shake the rags. The machine inclines on one side; and the rags being placed on the top, a large proportion of the loose dirt is beaten out by the time they reach the bottom.

(3.) *Rag Boilers.*—As the rags contain more or less grease, to get rid of it, and also to some extent to weaken the adherence of colouring matter, they are boiled with lime or soda in a boiler, which is made to revolve slowly round and round by the steam-engine,

thus causing the rags to be frequently turned over.
The old plan in Europe was to boil in a cast-iron vessel
by the direct action of the fire, and thus the pieces
which adhered to the sides were burnt. A great improvement was introducing steam for heating pipes
for the alkaline solution, and contrivances for filling
and emptying the rags, and making the boiler revolve.

(4.) *Washing Engines.*—After the rags are boiled,

they are torn or macerated until they become partly
reduced to half stuff or pulp. This is done in the rag-

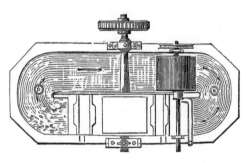

engines, consisting of iron cisterns, divided by a partition; a constant stream of pure water flows in at one end, and the dirty water passes off by a waste pipe. An iron roll furnished with knives revolves, and there is a *fixed* plate also at the bottom with knives. As the rags are drawn by the rapid rotation, they are cut by the knives to about one-sixteenth of an inch. The distance between the fixed and revolving knives is adjusted as required. There are other contrivances by which no material is allowed to be carried away with the dirty water, and for other purposes.

(5.) *Bleaching Vats.*—The pulp or half stuff is let down from the washing engines into vats built of brick,

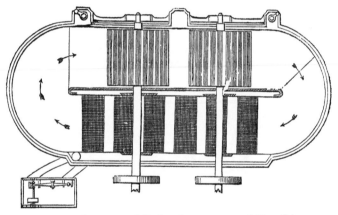

stone, or other material, for the purpose of bleaching or making white with chloride of lime. The quality which distinguishes the paper of the present day is its whiteness, which is obtained by bleaching; sometimes the bleaching solution is introduced into washing engines, and allowed to run about twenty minutes, then emptied to the bleaching department, where it remains for some hours. The stuff is then submitted to the action of an hydraulic press.

(6.) *Beating Engines.*—The pulp after being bleached

is placed in these engines, which are similar to the rag-
engines,—the only difference being that the cylinders
with teeth are made to move much faster, about 150
times a minute, so that the half stuff is reduced to a fine
pulp fit to be made into paper.   Supposing the revolving

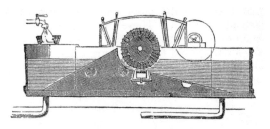

cylinder has eighty-four teeth revolving 150 times a
minute, and the bottom plate has twenty-one knives, in

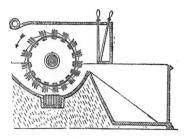

one minute it will make no less than 264,600 cuts.
Thus a large quantity of pulp is turned out in a few
hours, which could not be done by the rude stampers
used in India.   When colouring matter is required to
hide the imperfections in paper, it is introduced into
the beating engines.   Recently a centrifugal pulp
engine has been introduced to supersede the beating
engine.

(7.) *Paper making and Drying Machine.*—From the
beating engines the pulp when sufficiently ground runs

into large iron chests called *stuff chests*, in which an agitator is kept in motion for diluting the pulp with water. The stuff falls into a trough below, and passes through a brass plate pierced with very minute holes,

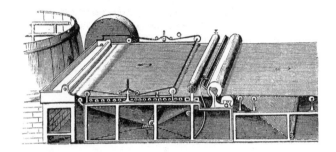

by which all knots or impurities are held back. The clear pulp is received upon an endless revolving wire cloth or gauze, so fine that in one square inch there are 3,500 to 5,000 holes. The wire gauge moves over a series of copper rollers to which a shaking motion is given by machinery, to facilitate the escape of water through the pores, and to distribute the pulp equally.

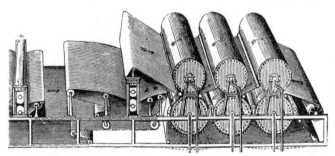

By means of air-pumps and other contrivances the pulp as it advances on the wire gauge solidifies, and passes onwards to a pair of rollers, the surfaces of which are kept wet by a jet of water, so that the pulp may not adhere to them. Then it is subjected to slight pressure, which is repeated by another pair of rollers; and at the same time a water-mark or any other design is impressed on it. The half-formed paper now passes upon an eldless felt, and receives on its way more pressure from other rollers, then passes through size solution, and is conducted over a series of large cylinders in which steam is employed to dry the paper. Thus liquid pulp is made into finished paper within three minutes in this machine.

(8.) *Paper Cutting Machine* is generally attached at the end of the paper making machine. As soon as the

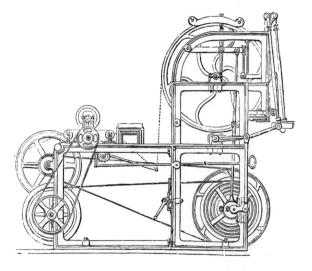

paper to be cut off descends from a drum it is taken hold of, and the length instantly cut off by a movable knife. After being cut, the paper slides down over a board in sheets, and is removed. The paper travels in the machine at about eighty feet per minute; it is temporarily arrested while being cut, and yet so beautiful is the contrivance employed in the machine that the paper is not creased in the slightest degree. There are paper bag machines also in which from a continuous length of paper the proper quantity is cut off, folded, pasted and delivered. From 1,000 to 4,000 bags per hour of a capacity to hold from two to twenty-eight lbs. can be thus made.

(9.) *Finishing.*—The cut paper is counted into quires of twenty-four sheets, and afterwards into reams of twenty quires; these are subjected to very heavy pressure in an hydraulic press, and then sent to market. The water-mark seen in some papers, and which is given to the paper when half formed, was originally adopted for the purpose of distinguishing the sorts and sizes of paper. The foolscap, the head-dress of a fool who in former times formed part of a great man's establishment, was a device of a later period, and a paper of a particular size is still called foolscap. The maker's name is now generally used, but in some cases more for ornament than distinction. The water-mark in bank notes and other documents is very important to detect fraud.

Varieties of Paper.—Manufactured paper is divided into many classes, but chiefly into three, *writing* paper, *printing* paper, and *brown* paper for wrapping. *Blotting* paper is not sized, and is rendered absorbent by the free use of woollen rags. *Parchment* paper is now made by immersing paper in a weak solution of sulphuric acid, quickly removed and washed off; this treatment produces great toughness, and gives the paper the appearance of parchment.

ESTIMATES FOR A PAPER MILL.

Machinery as shown on PLAN L.

£

Rag-chopper, or rope-cutting machine, with self-acting feed motion, set of knives, pulleys, complete, if required .   .   80

Rag-dusters and willows, with wrought-iron shaft, teeth, galvanised wire, wood casing, complete 125

2 Revolving rag-boilers, including all wheels, pulleys, steampipes, &c., to hold one ton of rags, complete   .   .   . 300

4 Washing engines, including cast-iron trough, with 54 steel bars in each roll, bottom plate with 14 steel bars, lifting and driving tackle, and fittings 650

For bleaching chests—the necessary fittings, and an hydraulic press, 10″ diameter, for half-stuff, with half-stuff boxes lined with copper   220

4 Beating-engines, including cast-iron trough, each roll with 84 steel bars, bottom plate with 21 steel bars, lighters fixed to lift from both sides, complete 750

Cast-iron girders and columns to support rag-engines and shafting, and also to serve as drains; pipes and valves to supply all rag-engines, and connections; gearing, including shafts, wheels, pulleys, driving-belts, &c.   .   .   . 700

1 Paper-making machine, to make paper 72 inches wide, including 2 iron stuff-chests 10″ diameter, 6 feet deep, with agitators, pulp-regulator, sand-catchers, brass knotter for brown and white papers. Wire part, including breast roll, 38 copper under tube-rolls, 4 copper rolls, copper dandy roller, brass deckle, apparatus for regulating width of paper and guiding 35 feet of wire, air-pumps and self-acting vacuum-boxes, save-all, pair of coucher or first-pressing rolls, pairs of metal first and second pressing-rolls, regulators for stretching and guiding felt, shake apparatus, with stands, pulleys, wheels, &c. Drying-machine, consisting of 10 cylinders 3′ 6″ diameter, fitted with steam and exhaust pipes, each cylinder to have a lifter to act either way, strong circular framing, cast-iron felt-rollers, copper leading-rollers to lead the paper round the machine. Sizing apparatus, including 2 brass rolls 8″ diameter, iron size-box, with fittings, complete. Calenders for smoothing paper, 4 rolls 12″ diameter, with all fittings, complete. All shafting, stands, pulleys, fixings for driving paper-machine, complete, with wheels for changing the speed of machine ; 1 revolving reel, complete, felts, &c.; 1 steam-engine, high-pressure, cylinder 12″, stroke 3′, for driving paper-machine, with arrangement for using waste steam for drying paper   1,900

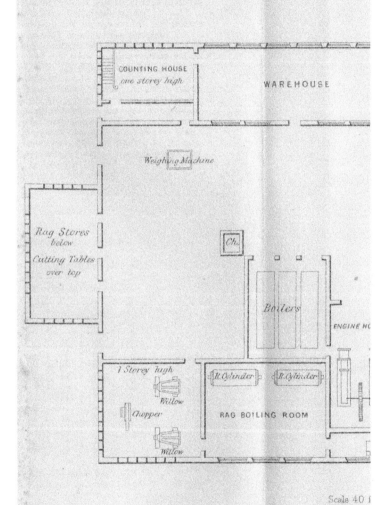

COUNTING HOUSE
one storey high

WAREHOUSE

Weighing Machine

Rag Stores
below

Cutting Tables
over top

Ch.

Boilers

ENGINE HO

1 Storey high

R.Cylinder

R.Cylinder

Willow

Chopper

RAG BOILING ROOM

Willow

Scale 40

0   5   10   20   30   40

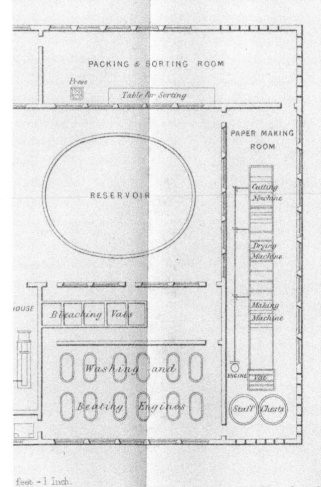

PACKING & SORTING ROOM

Press

Table for Sorting

PAPER MAKING
ROOM

RESERVOIR

Cutting
Machine

Drying
Machine

Making
Machine

HOUSE

Bleaching Vats

Washing and

Beating Engines

ENGINE

Vat

Stuff   Chests

feet = 1 Inch.

80

100 feet

Vincent Brooks, lith London.

£

1 Paper-cutting machine, longitudinal and transverse motions, cast-iron framing, revolving knife, pulleys, &c. . . . 130

1 Hydraulic press for finishing dry paper, ram 10″, including a set of hydraulic pumps with brass barrels, complete . . 165

Steam-engines, a pair, high-pressure and condensing combined, 50 horse-power, hot and cold water pumps, vacuum-gauge, governor, fly-wheel, with all connections, &c., plates for bottom of engine foundations, and for pump and condenser, &c., complete; 3 boilers of wrought-iron, with all the necessary mountings, injector, &c.; accessories and extras, including the necessary wire-work and knives for rag-cutting tables, rag hoist, water-cisterns with pumps to supply rag-engines, pipes for taking pulp to stuff-chests, extra articles for paper-making machine, iron rails with turntables for bleach-house, &c., boxes with wrought axles for carrying material, dry felts, wire, jackets for couchers, &c. 2,760

Packing and free delivery on board a vessel 778

£8,558

DIMENSIONS OF A PAPER MILL.

Machinery as shown on PLAN L.

Scale: 40 feet = 1 inch.

	Length, feet.	Breadth, feet.	Square feet.	Height, feet.	Cubic feet.
Rag store-room . . .	55	33	1,815	23	41,745
Rag-cleaning room . .	43	41	1,763	20	35,260
Rag-boiling room . .	52	41	2,132	12	25,584
Beating-engines room .	75	41	3,075	20	61,500
Bleaching-room . . .	75	23	1,725	10	17,250
Paper-making room .	134	33	4,422	12	53,064
Packing and finishing .	107	33	3,531	12	42,372
Warehouse	82	33	2,706	12	32,472
Offices, &c.	42	33	1,386	20	27,720
Engine-room	64	27	1,728	12	20,736
Boiler-room	40	35	1,400	10	14,000
Total of square feet and cubic feet			25,683		371,703

REMARKS ON PAPER MILL.

THE estimate includes all machinery for the general manufacture of printing, writing, and packing papers, to be made from rags, old gunny bagging, or any material likely to be used for such purpose.

Paper-making machines vary in width from 54 to 102 inches; and it is the width of the wire gauze of the paper machine which is the principal element for determining its capabilities. The estimate is given for one machine, which will make paper 72 inches wide; this is a medium size generally adopted in England.

Rags will answer very well both for printing and writing papers. From straw a thick brown paper is made in England; but for printing and writing only an inferior description can be produced, and of little comparative strength. The cost in England of producing two kinds of paper of equal quality, one from straw and the other from rags, is nearly equal. The waste, and the expense of preparing straw pulp for the production of paper, are so much more for chemicals, power, and labour, that in the end its cost nearly equals that of rags. In India and China it will be far cheaper to use rags, or any such material, for making paper, than straw or grass.

The waste in working rags in the several processes of dusting, washing, boiling, and reduction to half stuff, will be in proportion to the quality of rags. For very fine white rags about 10, for coarse 13, for coloured 18, and for old pack cloths and ropes from 20 to 30 per cent. is the approximate average of waste. In boiling rags, lime or soda is used, the choice depending entirely on its cheapness. By using revolving boilers there has been over the old method an economy of lime 50, of time 67, and of fuel 38 per cent.

The production of paper from the 72 inch machine will be 8 tons, more or less, per week, according to the speed at which the machine is driven for different qualities of paper. Paper mills in Europe generally work day and night. The paper mill near Calcutta, working with improved machinery, has made lately arrangements to work at night by gas made on their premises. Writing papers, highly finished, require great care in the preparation of pulp, and also a large amount of labour and machinery is required in finishing. It would be well, therefore, if attention is solely confined to produce paper of that quality for which there may be the greatest demand.

In sizing a certain class of printing papers, animal gelatine, made from the parings of bullock or buffalo hides, with a mixture of alum dissolved in it, is used, and placed in a trough between the drying cylinders of the paper-making machine.

Water for the manufacture of paper will be required in abundance, and that of a good quality : about 80 cubic feet per minute, besides the quantity required for condensing steam-engine. In the production of fine paper the quality of the water is very important ; and it is for this reason that paper mills in England are widely scattered, being generally built near pure streams of water. If the quality of water is such as to require filtering, two or three large ponds will be necessary to settle it.

In the paper mill near Calcutta, the water has first to be pumped from the river, and then settled in large ponds, and again pumped into the mill cistern; this has cost a good deal. Pumping apparatus will be required according to the depth of water to be raised into the tanks. If clear water could be obtained by a single pumping, the arrangement would be simpler, and much less expensive.

Water power may be employed in driving the machinery for preparing the half stuff for making paper, if the fall of water near a stream is sufficient, and to be relied on at all seasons, which is not generally the case. The paper-making machine is invariably driven by a separate small steam-engine attached to it, as it requires a perfectly regular motion; and the waste steam is employed in heating the cylinders for drying the paper.

Artificial parchment is also made in England from the parings of raw hides, by process of manufacture identical with that of a paper mill.

The cheapness of many foreign papers is due to a large proportion of white clay in them; some samples contain 30 per cent. and more, the presence of which is not apparent from external appearance.

The Plan L of the Paper Factory shows the rag store room apart by itself, separate from the main building; as the rags, from containing moisture and being overheated in a closed room, are liable to take fire. The store room for rags is below, and tables for rag-cutting above; from which room the rags are carried to the adjoining room for dusting, then to the boiling room, then to the rag engine room, which is one story high, in which there are columns and girders, included in the estimate.   This is a great improvement, as it serves for rag engines, for supporting shafting, for carrying off the dirty water, and also for drainage of the rag engine floor.   This arrangement will also save masonry work. Near the rag engine room, in the adjoining wing, are iron chests for half stuff, and the paper making and cutting machines ; next to it are the finishing and packing rooms, warehouse, and offices.

The cost of machinery and buildings, water tanks, &c., may be roughly calculated at from £17,500 to £20,000.

QUANTITY AND VALUE OF PAPER AND STATIONERY
EXPORTED FROM GREAT BRITAIN TO INDIA.*

Year.	Paper.	Value.	Stationery.
	cwts.	£	£
1866	27,335	91,713	31,777
1865	25,481	90,123	32,937
1864	30,642	105,634	32,867
1863	28,277	98,174	37,323
1862	30,403	104,945	31,220
1861	19,302	89,977	35,811
1860	20,187	96,752	35,622
	Including paper and stationery.		
	£		
1859	174,811		
1858	177,882		
1857	160,837		
1856	171,740		

* From Board of Trade Returns.

IRON FOUNDRY AND WORKSHOPS.

INTRODUCTION.

The commerce between Asia and England, and other parts of the world, is now conducted by a prodigious mercantile fleet. One single steam navigation company—the Peninsular and Oriental—possesses nearly fifty large steamers, a number equal to the navy of a second or third rate state in Europe; and some of them nearly 3,000 tons burthen. The Company has a capital of £3,300,000.

At Calcutta, and in the magnificent harbour of Bombay, may be seen at one time hundreds of vessels with the flags of all nations flying on their masts. For repairs of vessels, making good any breakage of machinery, and such other work as must be executed on the spot without any delay, the Peninsular and Oriental, and other steam navigation companies, have established iron foundries and iron workshops, fitted with improved machinery. But it will be scarcely credited that, in several ports, for instance in Bombay, where, besides the shipping, there are several cotton presses, cotton mills, and other industrial works, carried on with the aid of machinery, up to within a few years all the private work in repairs, or making good the breakages in machinery, was obliged to be executed in the Government *Gun Carriage Factory*, and was, of course, done at the leisure and convenience of the Government officials; though now there are two or three small private iron foundries and workshops, but not one worthy of the name.

It is in the iron foundry and the smithy that the iron is converted into several required forms, either by casting it in the foundry,—or in the smithy where the pig-iron is turned into malleable bar iron. Then it is shaped into different forms by various machines designed for the special work, either by cutting off the superfluous portions in the form of chips, or by hammering with a steam hammer, and producing a given form by shaping and joining all pieces together.

In machine tools, a good many improvements have been introduced in Europe from time to time. In England, Mr. Whitworth has aimed at nothing short of absolute constructive perfection, in which he has been so successful, that in other departments of machine-making his principles have been adopted. It was he who first appreciated the importance and value of a method of scraping by machinery for obtaining true surfaces;

seeing that by the old method the emery used for grinding became fixed in the pores of the metal, causing a rapid wear of the surface, which at one time impeded the progress of improvement in construction.

Along with true plane surface, the means of accurate measurement were thought by him of primary importance, and are now universally acknowledged. In the workshops of Messrs. Whitworth, in Manchester, now carried on by a Limited Liability Company, may be seen a small machine by which a difference in length or thickness of the one-millionth part of an inch may be detected! Mr. Whitworth also introduced a uniform system of screw threads, which, from its merit, has now become very generally adopted, and standard gauges of size are manufactured for use in all workshops. It was Mr. Whitworth who first adopted the principle of hollow casting for giving strength with a minimum of material.

Lathes are now made in Europe that would turn with ease a mass of fifty tons in weight, or as many feet in length; planing-machines that will bring to truth and flatness surfaces of 40 feet by 10 or 12 feet; boring-machines that will scoop out cylinders more than 12 feet diameter; slotting-machines that will gradually chop asunder a block of steel a yard thick; shears that will bite through a bar of forged iron a foot wide; and steam hammers of twenty tons and upwards, falling twelve feet, whose blow at a distance is felt as that of an earthquake. All the machines in the turning-shop are so completed and connected in a series that each in turn takes up the work from a previous one, and carries it another step towards completion; so that the attendant merely conveys the work delivered from one machine and places it in the next, finally receiving it

complete from the last machine. These several
machines for working in iron, brass, and other metals
are self-acting, work with great precision and extreme
accuracy, and are adapted to the nature of the work,
whereby the need of skill and dexterity in the workman is not so indispensable.

A self-acting fan is now an indispensable machine
in smithies and foundries, and is also in very general
use in several other manufacturing arts. Compared
with the blast of the old bellows worked by hand,
it is far superior, and gives a uniform stream of air.
By its means the smith heats his work with great precision. In the smithy the main pipes from the fan
are furnished with nozzle pipes of different diameters,
and the pressure of the blast varies. The fan is also
used in foundries, with a cupola furnace.

In India some people pride themselves upon execut-
ing all kinds of work with a simple tool. But this is
a mistake, and the more so if the merit of the work
should show a natural aptitude in the workman ; for it
is certain that if he has made good work with a bad
tool, he would make better with a good one. Let Asia
work with improved tools, and her progress in the arts
will be rapid. To work with a few primitive tools
when better tools exist is a loss of strength and resources. The machines made in England for working
in metals have not been surpassed in workmanship,
precision of action, or beauty of form, by those of any
other nation. It will be a good sign of the progress
of improvement if Asia should require of England for
some years to come machinery a thousand-fold more
than she does at present ; machinery which she is so
well able to supply, with her coal and iron mines so
fully developed, to the mutual benefit of both countries.

PRACTICAL PROCESSES IN IRON FOUNDRY AND MACHINE WORKSHOPS.

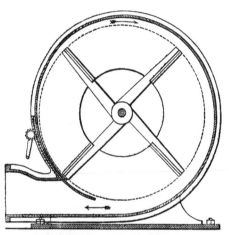

Iron Foundry.—The iron which is intended for castings is melted in a "cupola" furnace; the blast to the furnace being supplied by a self-acting fan worked from a steam-engine. When the iron is completely melted, the lower portion of the furnace is opened, and the white-hot stream of iron is carried in ladles to be poured into the mould. The mould is a box filled with sand, in which a wooden pattern of the intended casting is pressed, and the sand rammed down. A hole in the box receives the melted metal, which by a channel freely runs into the hollows left by the pattern, and completely fills it. When the metal has cooled, the casting is removed. If the casting to be made is not solid, but hollow, then solid metal patterns called *cores,* made a little smaller than the hollow left by the real pattern, are used.

Brass Foundry.—Brass is not a simple metal like iron, but is made by the direct mixture of copper and

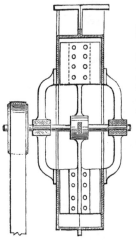

zinc. The brass furnace is built of fire-bricks. The crucible is formed of clay or plumbago. When the copper is melted, zinc is dropped in, in such proportions as may be necessary to obtain various degrees of hardness and colour. When the two metals are completely mixed, the crucible is withdrawn and its contents poured into moulds of sand. The method of making moulds for casting in brass is similar to that for iron. The materials for making moulds are fine sand or loam, and wood charcoal powder, which copies sharply all minute details of the pattern.

Self-acting Lathe is one of the principal cutting machine tools in a workshop, and is the oldest as well

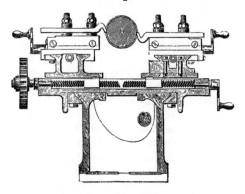

as the most serviceable. The slide-rest is an appendage

to the turning-lathe, and is so contrived as to hold a tool firmly to the work. The tool is held in a sort of iron hand or vice, which is made to move in the required direction by means of a slide by turning the handle; while the depth of the cut is regulated by an under slide. By the separate or combined action of the two slides the tool can be made to act along or across the work with great accuracy. Lathes are also made for ornamental turning, which are most complete self-acting tools for the production of regular forms, both simple and complex.

Self-acting Planing Machine is an application of the slide-rest for smoothing the surface, removing all irregularities, and producing a correct plane. The work is firmly bolted to the table, moving backwards and forwards under the cutting tool, which admits of accurate adjustment. Some machines are made to plane in both ways while the table advances and returns; but a great number are made to plane in one way, and a quick return motion is applied to the table to increase the performance and shorten the time. The cutting tool is kept cool during the work by allowing cold water to drip upon it, otherwise the edge would become soft.

Self-acting Slotting Machine is designed for the purpose of drilling slots or sinking slots through pieces of metal, or by cutting out of the solid metal. It is useful in replacing the labour of mechanics in chipping metal; it is further useful in economising the labour of smiths, as by its varied and rapid operations from a shapeless forging a finished form is produced, which by hand labour would consume much time, and would injure the quality of the metal by repeated operations. The efficiency of this slot-drilling is due to the form of the drill.

Self-acting Vertical Drilling and Boring Machine has its frame hollow, and for levelling and fixing work when too high for the table the foundation plate is planed and grooved. The table has a vertical slide adjustment, and swivels horizontally, to allow of larger objects being placed on the sole-plate of the machine.

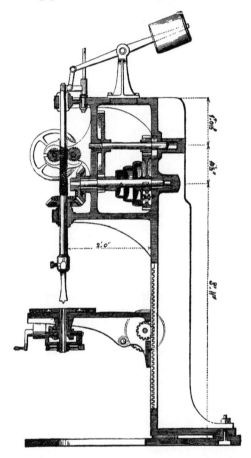

The machine is compact and self-contained. Smaller and similar machines are made, in which the driving gear is simple, and the table is raised or lowered by means of a screw and a handle.

Self-acting Radial Drilling Machine is made with a radial arm or jib carrying the drill, which is movable. It will compass any object within its range, according to the size of the machine. A means for turning the jib on its centre for adjustment or otherwise is provided in different ways by different makers. It is hung and swivels upon a bracket sliding vertically upon the flat surface of the frame to any required elevation.

Self-acting Shaping Machine is used for shaping levers, connecting rods, cranks, and similar articles that do not admit of being finished in the lathe. There are several self-acting motions for shaping horizontal, vertical, angular, and circular work, and also for hollow curves. For holding the work vertically or longitudinally there are tables. Shaping machines, like planing machines, cut one way, and are driven with a quick return motion, excepting for the smallest sizes. In some of these machines, instead of the work traversing under the drill, the drill travels over the work.

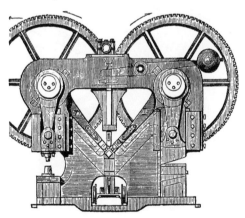

Punching and Shearing Machine is designed for punching holes and for shearing or cutting plates in the smithy and boiler shops. It has a triplicate lever applied for working the punch and shears, also for spacing out the rivet holes precisely to the length of the plate. The punch for making holes in the plates is balanced by a lever, and it may be raised or put out of action without stopping the machine. There is a contrivance for admitting long lengths of bars or narrow plates to be placed and cut off in the machine.

The Steam Hammer is capable of so gentle a work as cracking a nut, or of so tremendous a power as crushing a mass of iron. By means of a valve which is worked with ease and celerity, the velocity and force of the blow are regulated according as the nature of the work may require. The force of the blow may be varied instantly, from the utmost intensity to a mere touch, and any number of blows may be given with any rapidity up to 280 per minute. The hammer is worked with steam; as the hammer rises the steam is shut off.

ESTIMATES FOR AN IRON FOUNDRY AND MACHINE WORKSHOPS.

IRON FOUNDRY.

		£
1 Cupola fittings complete	.	230
1 Fan, 42″ complete	.	51
1 Grinding mill	.	250
1 Travelling crane, span 40	.	281
Miscellaneous, including a 3 ton ladle, 5 small ladles, 6 lifting chains, core carriage, stove door, &c.		206

SMITHY AND BOILER SHOP.

		£
6 Smiths' hearths	.	309
1 Fan, 22″	.	30
1 Plate bending machine, 6′	.	241
1 Punching and shearing machine	.	156
1 Steam hammer	.	195
Miscellaneous, grindstone complete, 6 sets smiths' tools and hammers, boiler makers' tools and hammers, anvils, blocks, stands, &c.		553

TURNING AND ERECTING SHOP.

		£
1 Slide break lathe, to take in 20 feet between centres .	.	471
1 Slide and screw-cutting lathe, bed 20 feet long . .	.	207
1 Slide screw-cutting lathe, bed 15 feet long . . .	.	156
1 Slide screw-cutting lathe, bed 12 feet long . . .	.	129
1 Hand-turning lathe, bed 16 feet long . . .	.	116
1 Hand-turning lathe, bed 14 feet long . . .	.	92
1 Hand-turning lathe, bed 12 feet long . . .	.	79
1 Planing machine to plane 12 by 4·6 by 4·6 . . .	.	423
1 Planing machine to plane 6 by 2·6 by 2·6 . . .	.	148
1 Vertical drilling machine to admit 3′·8″ diam., double geared		106
1 Vertical drilling machine to admit 3′·0″ diam., single geared .		65
1 Radial drilling machine, 8 feet radius	.	335
1 Radial drilling machine, 4 feet 6 in. radius . . .	.	132
1 Shaping machine, stroke 12½″	.	140
1 Shaping machine, stroke 9″	.	96
1 Slotting machine, stroke 18″	.	247
1 Slotting machine, stroke 12″	.	143
1 Vertical boring machine	.	333
1 Screwing machine	.	97
2 Grindstones with frames	.	75
Miscellaneous, including sets drills and steel tools required for the machine tools, chipping hammers, steel chisels, cylindrical gauges, and sundries	.	203
1 High-pressure steam-engine, boiler, gearing, shafting, pedestals, strapping, &c. Packing and free delivery on board a vessel	.	2,212

£8,507

ESTIMATE FOR A MECHANICS' SHOP TO BE ATTACHED TO MILLS AND FACTORIES.

1 12-inch double-geared sliding and screw-cutting lathe, self-acting bed 15 feet long, complete, with compound slide-rest, large face plate, shaft, pulleys, and hangers   190

1 Planing machine, self-acting, to plane 6 × 3 × 3, with quick return motion, complete   181

1 Vertical drilling machine, self-acting for holes 4 by 10 deep, with radial table, shaft, pulleys, complete   .   .   . } 81

1 Set taps and dies to screw ¼ to 1″, complete in case   .   .

1 Grinding-stone, set of steel drills, drilling tackle, and other tools and sundries   100

1 Steam-engine for driving all machines, complete; one self-acting fan worked by engine for smith's hearth, complete   .   260

£812

DIMENSIONS OF IRON WORKSHOPS.

Machinery as shown on Plan M.

Scale : 30 feet = 1 inch.

	Length, feet.	Breadth, feet.	Square feet.	Height, feet.	Cubic feet.
Iron foundry	93	44	4,092	40	163,680
Brass foundry, &c. . . .	30	23	690	20	13,800
Sand-mixing room . . .	18	12	216	10	2,160
Turning and erecting shop	134	54	7,236	30	217,080
Smithy and boiler-making shop	115	33	3,795	25	94,875
Iron store-room, pattern room, &c.	198	22	4,356	20	87,120
Total of square feet and cubic feet			20,385		578,715

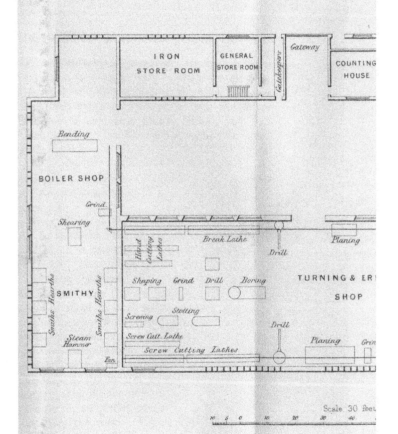

PLAN OF IRON

Scale 30 feet

M.

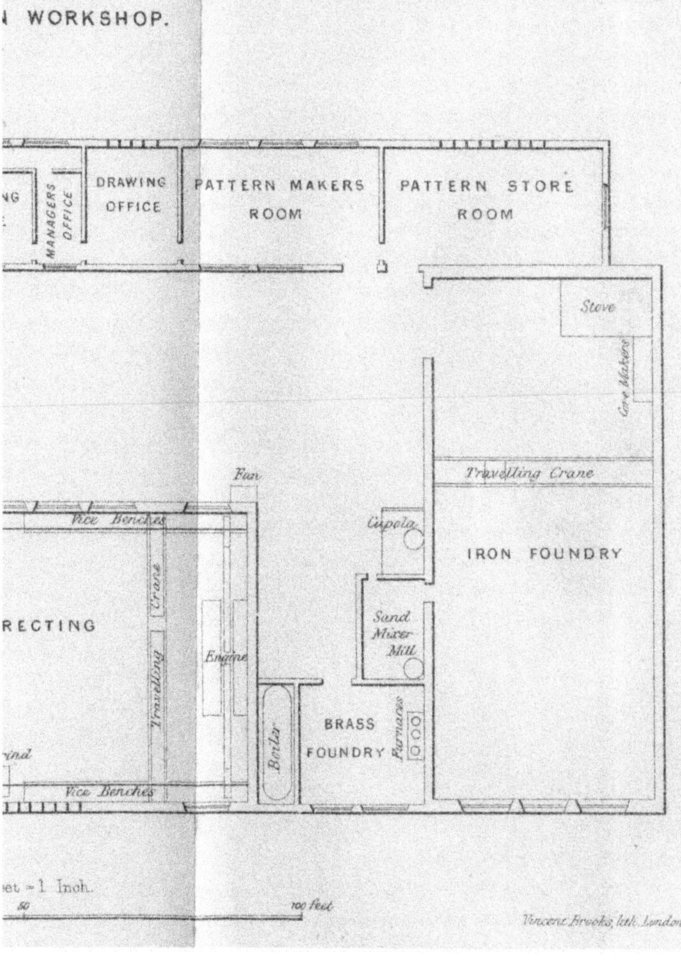
N WORKSHOP.
NG
MANAGERS OFFICE
DRAWING OFFICE
PATTERN MAKERS ROOM
PATTERN STORE ROOM
Stove
Core Makers
Travelling Crane
Fan
Vice Benches
Cupola
IRON FOUNDRY
RECTING
Crane
Travelling
Engine
Sand Mixer Mill
Boiler
BRASS FOUNDRY
Furnaces
Vice Benches
eet = 1 Inch.
50
100 feet

Vincent Brooks, lith. London.

REMARKS ON MACHINE WORKSHOPS.

THE estimate includes all machinery for iron foundry and workshops, on such a scale as to answer all purposes required in places visited by shipping, or where large industrial works are carried on with the aid of machinery—such as cotton presses, cotton, jute, and other factories, &c. Besides repairs, and making good any break-down in the machinery, the foundry workshops will be capable of turning out new steam-engines, boilers, and new machines of a moderate size, complete in every respect, and similar to those made in English establishments.

The Plan M shows the arrangement of the whole works. In the front wing, on one side, is the pattern shop, in which a facsimile of the article required is made in wood. Next to the pattern shop is the storeroom for these patterns. The iron foundry which adjoins receives these wooden patterns, and takes an impression of them in sand, running melted iron into the sand moulds, and thus producing the article in iron of the exact required form. As such casting of iron or brass is, comparatively speaking, rough and inaccurate, it is sent to the machine or turning shop, fitted with machine tools, to be turned, planed, and shaped. In the erecting shop all the different pieces of the whole work are fitted, completed, and made ready to be despatched from the works. This is the course of any piece of work executed in cast-iron. Wrought-iron, however, being a metal which cannot be cast, being malleable instead, is formed or wrought into the required shape by the aid of the forge and steam hammer, in the smithy-shop, which is placed on plan on the other side of the iron foundry. The turning shops are midway, so that all the subsequent operations of turning and planing in wrought-iron may be performed with equal facility as in cast-iron. Self-acting fans, driven by the steam-engine, have been placed in their proper positions for supplying blast to the furnaces in the iron foundry, and for the smiths' hearths. Both in the iron foundry and in the machine and erecting shop travelling cranes have been provided, so as to carry heavy articles with ease into any part of the shop.

Every machine in use in a workshop is constructed to perform a certain specific operation, and accordingly contains parts especially applied to the work in question ; which working parts are connected by the mechanism in such a manner that each shall move according to the nature of the work.

The total cost of the foundry and workshops, including all machine tools, as shown on Plan M, and specified in the estimate, including cost of buildings and other necessary expenses for putting the works in complete working order, may be put down at about £30,000. But workshops may be erected of any size, from £500 and upwards, and fitted up with machine tools designed for any special work, or for making any articles in large quantities that may be in general demand. With a view, therefore, of facilitating and determining approximately the cost of such workshops, the prices of the machine tools have been given in detail, so that any item may be omitted from the list, according to the amount of capital that may be proposed to be invested.

In a description of the extensive works of John Brown and Co., Sheffield, where large quantities of rails are manufactured for Indian railways, the "Engineering Journal," November, 1866, stated that "small quantities of Indian iron have been obtained, and it has proved fully equal to the best Swedish." The refined English pig iron is not sufficiently pure for conversion into steel, and large quantities of Swedish iron are imported into England for that purpose.

The machinery in some of the Calcutta jute factories has been made there, though a major portion has been sent over from England.

In every factory or mill established for any purpose, a mechanics' shop fitted up with tools, more or less in number, will be required, in proportion to the size of the factory, and the local facilities afforded for effecting repairs or making any new articles. A separate estimate for a mechanics' shop has been given, which will be applicable to a great many factories of moderate size. The best plan will be to drive the machine tools by a small portable engine, independently of the large engine in the factory.

In almost all towns in India, working in iron and brass is carried on for making domestic utensils, and for such other purposes, but by very rude and imperfect appliances, and entirely by manual labour. Thousands of hundredweights of yellow metal sheeting and sheet brass, besides large quantities of other metals, are imported into India. There is great room in some respects for introducing improved machines and appliances for working in metal, for articles in common every-day use, which are now produced by individual efforts or handicrafts all over India and other parts of Asia.

In Bombay, a native foreman employed at the gun carriage factory was the first to start a small iron foundry, and by honest industry succeeded beyond his expectations, and in a few years made a fortune. His example has been followed by two or three others, with complete success. But there is still a want in Bombay and other places of an iron foundry and workshops on a scale worthy to be called by that name. Calcutta seems to be better off in this respect, chiefly on account of the large numbers of Europeans in the Bengal Presidency; and more by their enterprise than that of the natives, the iron foundries and workshops in Calcutta have been established.

Two English gentlemen connected with the iron trade, who went to the Continent for the express purpose of making inquiries on the subject of foreign competition, wrote to the London *Times*, December, 1866, that "Belgium and France have thrust us out of the foreign markets. They have almost monopolised in Russia the trade in iron for railway purposes. They make the rails, they supply the locomotive engines, the roofs for stations, and pillars, and they build the carriages. A like state of things obtains in Spain. Even at home in England, within our own boundaries, these pushing people are challenging our supremacy."

VALUE OF RAW METALS IMPORTED INTO INDIA FROM GREAT BRITAIN.*

Year.	Copper.	Zinc.	Tin.	Iron.	Steel.
	£	£	£	£	£
1865	1,661,987	101,887	255,662	586,712	60,943
1864	1,361,416	119,863	151,924	724,706	40,963
1863	1,169,641	97,190	104,384	678,312	80,129
1862	991,098	110,418	130,065	603,222	96,773
1861	1,082,312	120,527	121,763	454,438	45,617
1860	820,135	226,035	95,352	571,839	53,967
1859	651,947	88,315	64,885	1,107,222	73,110
1858	369,049	45,993	88,255	494,054	51,680
1857	387,431	56,974	37,208	464,576	33,517
1856	322,485	60,738	66,009	222,261	29,871
1855	260,220	47,961	63,434	218,674	30,446
1854	276,242	9,475	46,044	102,103	15,992
1853	210,230	11,303	77,868	145,248	10,216
1852	441,412	41,532	71,638	314,398	12,935
1851	878,923	78,994	68,673	464,649	14,315

VALUE OF MANUFACTURED METALS AND MACHINERY IMPORTED INTO INDIA FROM GREAT BRITAIN.†

Year.	Manufactured Metals.	Machinery.	Year.	Manufactured Metals.	Machinery.
	£	£		£	£
1865	608,104	554,156	1857	558,329	244,443
1864	418,673	585,516	1856	788,859	435,512
1863	424,188	506,518	1855	312,304	126,303
1862	383,694	553,883	1854	286,671	52,788
1861	386,748	870,251	1853	217,817	28,457
1860	454,457	871,531	1852	246,701	14,337
1859	447,011	587,566	1851	245,393	20,666
1858	378,989	465,453	1850	166,139	8,079

* From Statistical Abstracts of Colonial and other Possessions.
† From Statistical Abstract presented to Parliament 1867.

TIMBER MILLS

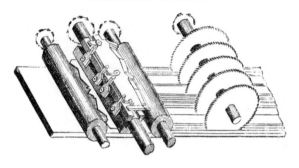

INTRODUCTION.

EVERY one is more or less interested in timber, so
largely is it used in building houses, factories, boats,
ships, furniture, and for other like purposes, and also as
fuel in many countries. Though a saw-mill for cutting
logs of timber, moved by water power, was in use in
Europe in the thirteenth century, on the Malabar coast
and in Bombay machinery for cutting logs of timber
was not introduced and worked successfully until within
the last six years. It was introduced, not by any native,
but by a European firm fully engaged in the cotton
and other piece goods trade. There is a general complaint, particularly in Bombay, that house rent is excessive and living very dear; and this is a matter of
no surprise, when we consider that logs of timber used
in buildings are brought to the place where the buildings are to be erected, and sawn by the tedious process
of hand-labour, and that the firewood is cut by hand

up to this moment in Bombay, a city which has more population than either Liverpool or Manchester. In England, in the smallest town even in the country, there is always a yard stored with timber, cut by machinery, of all sizes used in the building trade, well seasoned; and a house or factory can be built in less than one-fourth the time taken in India. For preparing all parts of doors and windows, machinery is employed; and large factories have been erected in some parts of Europe for turning out doors and windows for export to London and other places; while the same work is done by hand in India. For this reason, in some of the Bombay mills built latterly may be seen hundreds of *iron* windows made at Birmingham, which have cost less than timber windows would have cost. As in iron work, so in timber work, there are machine tools designed for the special work intended to be performed. Machines are made and used extensively for sawing timber, planing, edging, tonguing, grooving, and moulding, by which the production of work has been increased, and the price cheapened.

TIMBER WORKING MACHINERY.

Timber Frames are designed for sawing round or square logs of any description of timber into boards or planks of any required thickness. They are fitted with upright saws, and the timber to be cut is placed on an iron rack carriage with quick motion for running forward or backward, the speed of which is regulated with any degree of nicety. Adjustable clips for holding the timber whilst being sawn are attached to the carriage, for the purpose of following the irregularities of the logs. These machines are made to take in a log of timber from 50 to 20 feet long, and to saw from 42 to 16 inches. They carry one saw to every inch in width of the machine, according to the size it is made, and require from 10 to 3 horse-power to drive them. There are also *circular* saw benches made for cutting timber, plain or self-acting, with travelling bed.

Roller Planing Machine will plane, joint, tongue, and groove all kinds of timber, effecting all the processes at one operation, or each singly. The timber is propelled through the machine by means of rollers, some of which are so arranged that they can be raised or lowered together for various thicknesses of timber, by turning a wheel at the end of the machine. The under side is planed by two fixed cutters, which are easily taken out for sharpening. A revolving cutter is placed before the stationary cutters for the purpose of taking off the rough surface, and the edges are either jointed, rebated, or tongued and grooved, by means of cutters fixed in two revolving blocks.

Moulding Machine is adapted for cutting single or double mouldings of any pattern, and giving a smooth finish to the mouldings. The boards are brought forward by revolving feed-rollers, and the action of moulding is done on all its surfaces at once by means of four sets of revolving blocks carrying cutters of the shapes required, running at great speed. The mouldings are cut with great accuracy. Boards are also planed in this machine by attaching plain or straight cutters to the top and bottom blocks, and edged according to the kind of cutters fixed to the side blocks. The boards pass through this machine for moulding at the rate of from 10 to 30 feet per minute.

Circular Moulding Machine is adapted for moulding circles and wood of any irregular shape, such as circular heads of sashes, hand-rails, table edges, and various things in joinery and cabinet work.

Veneer Sawing Machine is for the purpose of cutting various kinds of hard and fancy woods into slices or veneers. The saws are extremely thin, and are attached in segments to a cast-iron disc, which is truly turned and balanced.

Cross Cut Saw Bench is intended for cross cutting planks to any required length, and squaring all the ends to the greatest nicety.

Tenoning Machine is designed for cutting the end of a piece of wood so as to fit into another piece, and is applicable for doors, windows, sashes, shutters, and other joiner's or cabinet work. The double tenoning machine is particularly applicable for cutting double tenons in framing for railway carriages, wagons, and other heavy descriptions of timber.

Vertical Boring Machine is made for cutting holes in wood, and will bore holes of any size up to 3 inches diameter, and 16 inches deep. It requires about one horse power to drive it. In some machines the cutting tool operates vertically, and in other machines horizontally. In machines of this class some are made in which a round hole is bored, and then other tools follow for squaring out the four corners and sides.

ESTIMATE FOR A TIMBER MILL.

£

1 Timber frame, 36 inches, for cutting logs or trees into boards or scantlings, with rack-carriage, worked by a pinion, with quick motion, clips, and saws, each buckled, complete 350

1 Planing machine, for working timber up to 5 × 12, and for jointing, tonguing, and grooving, either in one compound operation or singly; including assorted irons of different sizes 298

1 Moulding machine, capable of moulding up to 3 × 9; calendar rollers, including 1 set of cutters, complete . . 117

1 Veneer sawing machine, with 10-feet disk, sliding carriage, with rack and pinion, setting-in motion for the required thickness of veneer, including 1 set of segments, complete 231

1 Plain saw bench, to carry a saw 42 inches diameter, with spindle and pulley, including 2 saws, 36 and 42″, complete . 50

1 Steam - engine, horizontal, high - pressure, diameter of cylinder 16″, stroke 30″, fitted with piston, governor, improved valve, water heater, &c., complete; 1 boiler, 20 × 5·6, fitted with safety-valves, water and steam-gauges, &c., complete; shafting and gearing for driving all machinery, including all wheels, couplings, pulleys, pedestals, columns, lubricators 711

Miscellaneous, including moulding grinder, grinding machine, belts for driving, extras, and sundries; and packing, and free delivery on board a vessel 332

£2,089

ESTIMATE FOR WINDOW AND DOOR MACHINERY.

£

1 Squaring machine, for planing and squaring timber for the framing of doors, sashes, &c., to plane 18 feet long by 18 inches wide, with disk and cutters, travelling-bed, including cutters, complete 157

1 Cross-cut saw bench, self-acting, arranged for different thicknesses in wood, including 2 saws 35

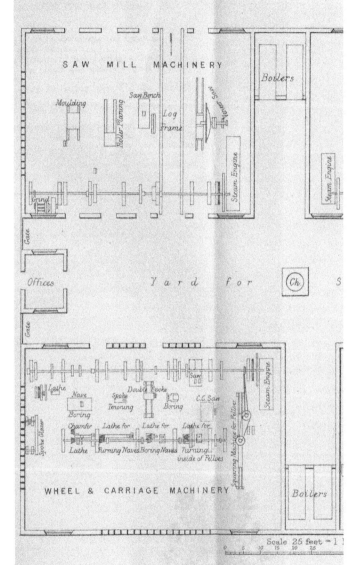

PLAN OF SAW MI

SAW MILL MACHINERY

Boilers

Moulding

Roller Planing

Saw Bench

Log Frame

Cross Cut Saw

Steam Engine

Steam Engine

Grind

Gate

Offices

Gate

Yard for S

Ch

WHEEL & CARRIAGE MACHINERY

Lathe

Spoke Planer

Nave Boring

Chamfer

Lathe

Spoke Tenoning

Double Spoke Boring

Lathe for Turning Naves

Lathe for Boring Naves

Saw

C.C. Saw

Lathe for Turning inside of Fellows

Squaring Machine for Fellows

Steam Engine

Boilers

Scale 25 feet = 1 I

0   5   10   15   20   25

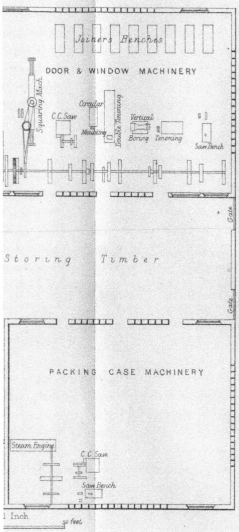

Joiners Benches

DOOR & WINDOW MACHINERY

Squaring Mach.

Circular

C.C.Saw

Moulding

Double Tenoning

Vertical

Boring   Tenoning

Saw Bench

Storing   Timber

Gate

Gate

PACKING   CASE   MACHINERY

Steam Engine

C.C.Saw

Saw Bench

1 Inch   50 feet

£

1 Circular moulding machine, for circles and irregular shapes, a double-edged cutter, complete   38
2 Double tenoning machines, bed having a lateral slide, cutter, with rack and pinion and balance-weights, including 6 cutters, complete   225
1 Vertical boring machine, for boring holes 3″ diam., 16″ deep, including 6 augurs   46
1 Saw bench, for sawing, tonguing, &c., parallel fence, &c., complete   33
1 Steam-engine, horizontal, high-pressure, cylinder 13″ diameter, stroke 24″, complete; 1 boiler, complete, with all fittings, valves, pipes, &c.; shafting and gearing, including all wheels, pedestals, pulleys, iron columns, &c. . . .   679
Miscellaneous, including belts for driving and other sundries, packing, including free delivery on board a vessel . .   247

£1,460

ESTIMATE FOR PACKING-CASE MACHINERY.

£

1 Cross-cut saw bench, complete, including 2 saws, 16″ and 18″ .   35
1 Plain saw bench, complete, including 2 circular saws, 18″ and 24″   20
1 Steam-engine, high-pressure, boiler, shafting, belts, sundries, extras, packing, delivery.   215

£270

DIMENSIONS OF A SAW MILL.

Machinery as shown on PLAN N.

Scale, 25 feet = 1 inch.

	Length, feet.	Breadth, feet.	Square feet.	Height, feet.	Cubic feet.
Saw mill, for cutting logs of timber . . .	65	50	3,250	12	39,000
Door and window-making room	74	50	3,700	12	44,400
Packing-case room . .	65	50	3,250	12	39,000
Wheel-making room .	74	50	3,700	12	44,400
2 Boiler houses . . .	18	15	270	10	2,700
2 Boiler houses · . .	18	15	270	10	2,700
Total of square feet and cubic feet			14,440		172,200
Open yard for storing timber, 150 by 40 feet.					

REMARKS ON TIMBER MILL.

MACHINERY for working in wood is made suitable for builders, joiners, contractors, and timber merchants, each adapted for its special trade, either for temporary or permanent works of all sizes, and driven by a portable or a fixed steam-engine.

The estimates for the timber saw-mill and other machinery are given for a complete permanent work, and include all machines for sawing, planing, and working the same in doors, windows, mouldings, &c., and on a scale to answer all the requirements of a complete timber mill.

With a view of making the estimates useful and applicable in more instances than would be the case otherwise, they have been divided into four sets, as follows, and the cost of machinery in round numbers for each will be for the—

	£
Timber saw-mill . . .	2,100
Window and door machinery . .	1,460
Packing-case machinery . .	270
Wheel-making machinery . .	2,334

—including steam-engines and boilers so made as to burn wood shavings and waste of the factory, and requiring a proportionably less quantity of coals.

The total cost for the entire works complete for carrying on all the operations at one establishment, and answering all the purposes of a complete timber factory, may be put down—including buildings, and erection of machinery in working order—at from £13,000 to £16,000. However, in the estimates the price of each machine, the use of which has been defined, is given separately; so that machinery for working in timber may be erected on any scale, and any items omitted from the estimates which may not be required. Thus timber mills may be erected of any size, and from £500 and upwards, and fitted with machines adapted for any special work.

The Plans N of timber mill have been so arranged that each of the four blocks may be erected separately, or all together as one establishment, where all the operations may be carried on one after the other in the same works, and in the same place. Thus each plan is complete in itself, taken separately, with its steam-engine and boiler, or the four may be combined together as one, as shown on Plan N.

CORN MILLS.

PROVIDENCE has scattered grain, in many of its varieties, widely over the earth. Rice is extensively cultivated in Asia, from whence it has extended gradually to the southern parts of Europe, and has been introduced also into America. Wheat and barley extend to all the temperate regions.

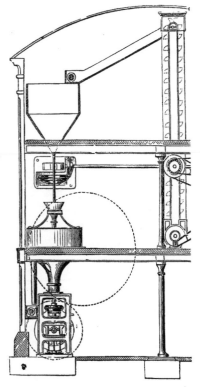

For the food of man it is well understood that no article is so well adapted as corn in many of its varieties, as it contains all the ingredients for forming and sustaining the fat, the muscle, and the bone in the human body. Throughout all parts of Asia it is used as the principal food of the inhabitants, more so than in Europe, though even in Europe bread is called the staff of life.

In such a simple art as that of reducing the grain to a powder, or to a state fit for making bread, the Eastern world is many ages behind the Western. The art of grinding corn into flour in Asia is practised up to the present time as it was thousands of years ago. It is the practice for the women to grind corn by means of a hand-mill, which is to be found in every house—rich or poor. The hand-mill consists of two flat stones about two feet in diameter, kept rolling one on the other by means of a stick. The corn falls down through a hole in the middle, and by the circular motion it is bruised and reduced into flour, which falls out at the rim of the millstones. Such is the primitive way in

which the grinding of some of the most important articles of food has gone on in Asia from generation to generation, without any improvement up to the present time.

In Europe, the art of grinding corn has undergone many modifications. Corn mills were erected, in which at first wind or water power was made use of in driving the stones. But it was only at the close of the last century that important progress was made in the process of grinding, or in the development of those principles by which the whole operation is now reduced to a system. By the introduction of steam into corn mills, as into every other manufacture, a great change has been effected, giving certainty and effect, and reducing the cost of working. Corn mills are made to depend no longer upon the state of wind or the supply of water, the whole operation being now carried on by steam power. Mills are erected in the very heart of European towns, and most of the changes and improvements visible are owing to the use of steam.

In a modern corn mill little manual labour is required. The corn to be ground is conducted by a self-acting elevator to a machine, where it is cleansed from the dust and other extraneous matters which are found more or less combined with it, by the action of a blast from a self-acting fan, and the dust is carried off through a closed passage to the outside of the mill. The grain being thus cleansed is greatly improved in whiteness and wholesomeness. It is then taken up by another self-acting elevator, and by an endless screw-conveyor delivered into the opening or hopper which surmounts the millstones. The stones revolve rapidly like other machines by wheels connected with the steam-engine. In the hopper which supplies corn to the millstones, a jigging motion is kept up, so as to shake the corn over the stones in equable quantities, and thus a regular supply enters between the stones. So long as this action is going on properly a little bell is made to ring, the motion of which ceases with the supply of wheat. The stones are completely boxed in, and a blast of air plays upon the grinding surfaces of the stones for the purpose of keeping them cool and removing the flour as fast as it is formed ; the effect of which is, that nearly twice as much corn can be ground per hour as on the old system, and when the machinery is in operation, no dust or flour is flying in the mill.

The corn, after being reduced by the millstones into flour, is taken up by a revolving band fitted with a number of leathern cups to a dressing machine, covered with wire or silk cloth of different degrees of fineness ; in which the meal is separated into fine and coarse flour of different qualities, and bran. The finest of the flour passes through the upper end of the machine which contains the finest wire or silk cloth ; the next division, fitted with coarser cloth ; and so on, producing different qualities of flour ; while the bran, which is the outer husk of the grain, and used for feeding cattle, being too coarse to go through any of the divisions of the wire cloth, is discharged at the other end of the machine.

It is a melancholy fact, that in the vast Eastern Continent, producing every variety of the principal cereals, and where corn forms the principal article of food of the majority of the people, and is so essential to the maintenance of health, there are so few, hardly any, corn mills working with improved modern appliances.

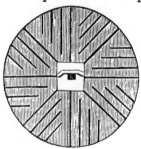

A CORN MILL.

(1.) The *Smut Machine* is used for the purpose of cleaning all kinds of grain before grinding, and separating dust and small gravel. The grain, in the first place, is elevated from a reservoir in the grain store-room by a self-acting apparatus without manual labour, and fed into this cleaning machine at its upper extremity. Then it is thoroughly agitated in its passage through this machine, in a wire cage, and exposed to a current of blast from a fan, which completes the removal of chaff and dirt. The result is that the whiteness of the flour is greatly improved, and also its wholesomeness, on account of the impurities being separated, which collect in the casing.

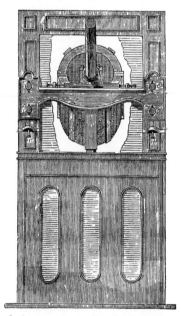

(2.) *Grinding Stones.*—The cleansed corn is elevated again, and by an endless screw conveyor conveyed into the hoppers of the grinding stones, where by machinery a jigging motion is kept up, so as to shake the corn, and let it fall over the stones in equal quantities. So long as this action is going on properly a

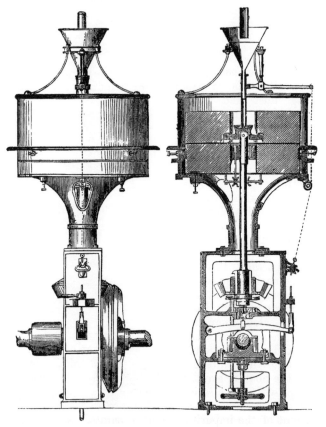

little bell is made to ring, the motion of which ceases with the supply of wheat. The stones are so dressed, grooved, and divided into parts, that when working the sharp edges meet each other like a pair of scissors, and grind corn effectively. In modern mills an artificial blast is also used to keep the stones cool at the time of grinding.

(3.) *Dressing Machine.*—Another elevator carries the ground grain, or meal, from the stones to a dressing machine above the stones, consisting of a revolving hollow cylinder placed in a slanting position, turning about 650 times a minute. The cylinder is covered with fine wire gauze, or fine silk cloth made for the

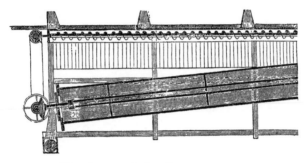

purpose. The finest of the flour passes to the upper end of the dressing machine, which contains the finest cloth, then to the next division of coarser cloth, and so on, till the bran, which is too coarse to pass through any of the meshes, falls out at the other end of the machine. To the bottom of the machine are attached sacks, which, being filled with clean flour, are removed as required.

R

ESTIMATE FOR A CORN MILL.

Machinery as shown on PLAN O.

£

1 Corn elevator for conveying corn over top of smut machine ;
  1 smut or corn-clearing machine, complete, with all fittings
  1 elevator for conveying cleaned grain from smut machine
  to top of mill stones, and screw conveyor for conducting
  wheat over stone hoppers 88
3 Pairs grinding stones, 48 inches diameter, best, with casings,
  feeding and disengaging gear, with all fittings complete,
  with endless conveyor 332
1 Flour elevator for conveying meal from bottom of stones to
  1 double dressing machine, with silks, complete; 1 separa-
  tor for the offals, with rotating sifter, complete . . . 110
1 High-pressure steam-engine, horizontal, 16 H. P., and 1 boiler,
  doubled flued, with all fittings complete 412
Mill gearing for stones, upper and lower, wrought-iron shafting,
  wheels, pulleys, columns, &c. 124
Miscellaneous, including fan and exhaust apparatus, driving bands,
  crane for raising runner stones, sacking cylinders with valves,
  &c. ; packing and free delivery on board 110

£1,176

DIMENSIONS OF A CORN MILL.

Machinery as shown on PLAN O.

Scale: 12 feet = 1 inch.

	Length, feet.	Breadth, feet.	Square feet.	Height, feet.	Cubic feet.
Room with upper story containing stones and other machinery	36	29	1,044	26	27,144
Store-room for grain, shown on plan *without* scale .	18	29	522	26	13,572
Store-room for flour, not shown on plan . . .	36	19	684	11	7,524
Engine and boiler house .	25	29	725	13	9,425
Total of square feet and cubic feet			2,975		57,665

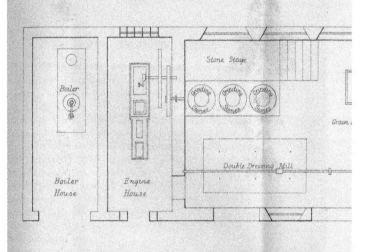

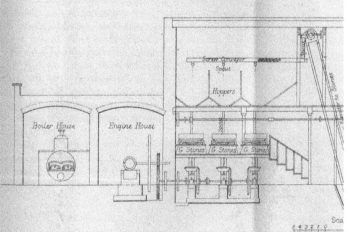

Scale

OF A CORN MILL. O.

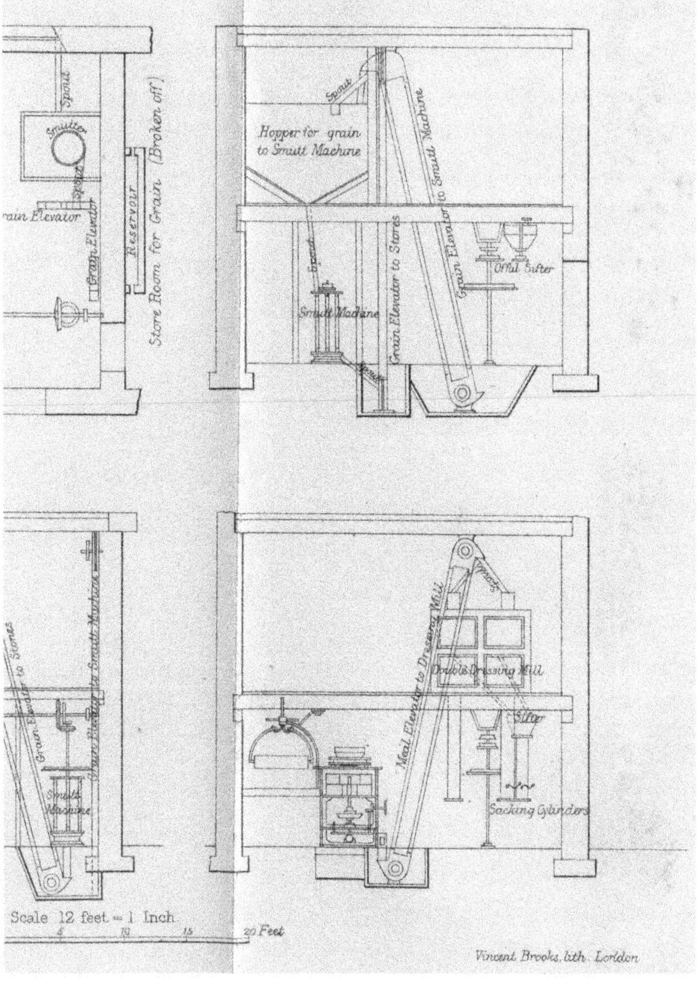
Spout
Smutter
Spout
rain Elevator
Grain Elevator
Reservoir
Store Room for Grain (Broken off)
Spout
Hopper for grain to Smutt Machine
Grain Elevator to Smutt Machine
Grain Elevator to Stores
Spout
Smutt Machine
Offal Sifter
Grain Elevator to Stores
Grain Elevator to Smutt Machine
Smutt Machine
Spout
Meal Elevator to Dressing Mill
Double Dressing Mill
Sifter
Sacking Cylinders
Scale 12 feet = 1 Inch
5
10
15
20 Feet

Vincent Brooks, lith. London

REMARKS ON CORN MILLS.

PLAN O shows the arrangement of the corn mill machinery so adapted as to make it work with as little manual labour as possible. The corn will be elevated by a self-acting mechanism from a reservoir in the store-room. Then it will be conveyed automatically from one machine to another, ground into flour, and the different qualities of fine and coarse flour separated, and the sacks filled up ready to be removed to the warehouse. The elevation and sectional plans will show distinctly how this is done, and the way in which each machine is driven. The store-room for grain is not drawn on scale, and the warehouse for flour is not shown, but the sizes of both have been given in the dimensions of plan.

The estimate for the corn mill provides for three pairs of stones, with all necessary machinery for working on the modern system, and adapted for grinding rice, wheat, and all other kinds of grain.

In most English mills the diameter of the mill-stones is 4 feet, and their thickness about 12 inches. For wheat one stone is composed of French burr, which is very hard but porous. As it is difficult to get the stones of the required size it is usual to construct them in segments, firmly bound by iron hoops round the circumference.

In the estimate the cost of a dressing machine with fine silks for separating the finest flour is included. If a dressing machine with iron wire be substituted it will cost about £60 less. According to the fineness of flour required, instructions should be given for the precise kind of dressing machine.

The total cost as per estimate for the three grinding stones, with other necessary machinery, and including packing and free delivery, amounts to little more than £1,000.

The production of flour from the mill for which the estimate is given will be 150 bushels of wheat every ten hours. It has been stated that in many corn mills in England, fitted with the most improved machinery, each pair of stones grinds even more, of coarse flour.

During the siege of Sebastopol a steam vessel was fitted up with *four* pairs of grinding stones, and other necessary machinery, for grinding corn for the use of the English army. From the official reports giving results of the working of this floating mill, it appears that when in harbour the daily produce of flour of a quality fit for the army was 24,000 lbs., which was more than had been calculated on; and the expenses of working, including the establishment of the vessel, were 3s. 1d. per 100 lbs. of wheat ground. During a certain period the total quantity of wheat supplied to be ground was 1,776,780 lbs., from which the quantity of flour was 1,331,792 lbs., with 358,172 lbs. bran.

Portable grinding mills, for grinding from 6 to 10 bushels, driven by a portable steam-engine, capable of grinding every shade of fineness, complete, with stones, are constructed, and may be had from £100 upwards.

EXPORT OF RICE, WHEAT, AND OTHER GRAIN FROM
INDIA TO GREAT BRITAIN.*

Year.	Rice.	Wheat.	Other Kinds.	Quantity, Rice.
	£	£	£	Qrs.
1866				
1865	5,573,537	110,265	272,606	
1864	3,975,565	78,676	271,136	
1863	3,378,496	112,056	237,358	2,201,183
1862	3,635,075	147,501	257,362	4,576,553
1861	2,962,497	135,059	253,321	3,795,137
1860	2,276,296	112,222	200,044	2,233,864
1859	2,433,145	116,945	251,781	2,004,667
1858	3,449,172	142,761	198,441	
1857	2,301,182	138,499	147,775	
1856	2,598,070	173,883	124,309	
1855	1,562,318		180,212	
1854	1,261,503		152,151	2,474,399
1853			889,160	1,158,753
1852	Value of rice, wheat,		869,002	921,555
1851	and other grain in one		752,295	771,572
1850	sum.		757,917	818,992

* From Statistical Abstracts presented to Parliament, 1867.

BRICK WORKS.

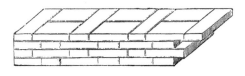

INTRODUCTION.

THE art of making bricks is of the highest antiquity.
Throughout the greater part of the world, according to
the facilities of the soil and the necessities of the
climate, it has been more or less successfully practised.
In every country, with the ordinary progress and
requirements of increasing population, commerce, and
wealth, brick buildings are superseding the temporary
buildings of wood and mud which yet prevail in several
regions.

Although the art of making bricks is very ancient,
it scarcely received any improvement from the aid of
machinery until within the last few years. Machinery
is now employed for grinding and crushing the clay,
and for making bricks. The bricks made by hand are
generally of very inferior quality, while those made by
machinery, in appearance and strength, and adaptability
to architectural purposes, are far superior. By the use
of machinery the great inconveniences attendant on
moulding the bricks by hand, and the delay in drying
the hand-moulded bricks, caused by the presence of
excess of water in the clay, are also obviated to a great
extent. When the process is carried on with good

machinery in a factory, the whole goes on in harmonious order, with ease and regularity; and the rapidity and amount of production in a given time by machinery are also great when compared with the hand process. The bricks are more compact, more perfect in form, and the edges more sharply defined, and so smooth and sightly in appearance that for facing the outside of houses no plaster is at all required. Ventilating bricks and hollow bricks are now extensively made in England, which take less material and are lighter, as well as bricks adapted for partition walls, for arches, doors, chimneys, &c.

The machinery for making bricks is divided into two classes; the one operates on the clay in a moist and plastic state, while the other requires it to be dried and ground previous to being moulded. In the former class the plastic clay is formed into a continuous length in the machine, and is then divided into bricks by means of wires moved across by manual labour. This process requires the clay to be soft. The bricks thus made are only a little harder than those made by hand, and require to be dried before being placed in the kiln. In the second class, the dry process, the bricks are compressed in a dry state in the mould; but the process of drying the clay and reducing it to a uniform powder requires more motive power to drive the machinery.

It has been stated on very good authority that the first requisite for the proper working of wire-cutting machines by the wet process is, that the clay should be a strong plastic kind, or it will not pass through the orifice or die of the machine with a surface sufficiently smooth for the face of the brick. With clay of this degree of plas-

ticity, the drying and burning processes, so important
to the manufacture of sound bricks, are apt to be
irregular, and to cause a large proportion of unsound
and inferior make to result. In order that the bricks
may be dried and burnt in the kiln in a regular and
sound manner, it is found necessary to reduce the te-
nacity of the clay by the mixture of sand, &c., in certain
proportions, so that the bricks may not crack during
the operations of drying and burning. But if the sand
be not finely divided and well mixed in the mass it is
apt to impede the cut of the wires, and to leave rough
uneven ends. For these and other reasons the dry
process of manufacturing bricks is more generally
adopted.

As regards burning the bricks after being made by
either class of machinery, Professor Thomson, in a
paper read before the Chemico-Agricultural Society,
pointed out the great loss of heat which occurs in the
ordinary modes of burning in the common kilns; the
air which has passed through the fuel or among the
heated bricks, and the smoke, pass away from the kiln
to waste at a very high temperature—even at a red
heat—during a considerable part of the process. When
the bricks are raised to the high temperature required
to burn them, and render them permanently hard, the
great store of heat which they contain is entirely
thrown to waste while they cool. He stated that he
has noticed with much interest the very admirable
principles of a new kind of kiln with perpetually
revolving fire, invented and patented by F. Hoffmann,
of Berlin, both for brick-burning and lime-burning.
In this new kiln economy of fuel is effected in a twofold way; in fact, by saving the twofold loss of heat;

for, first, it saves the heat of the gaseous products of combustion and unconsumed air passing through and away from the burning bricks, by applying this heat effectively in drying the new fresh bricks about to be burnt, and raising them up to an incandescent temperature, so that only a very slight addition of heat directly from ignited fuel is required to complete their burning; and, secondly, it saves the heat of the cooling bricks, after their having been sufficiently fired, by applying it all in warming the air which goes forward to supply the fires; so that the fuel is burnt with air already at an incandescent temperature, instead of requiring, as usual, to heat the air for its own combustion.

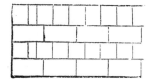

MANUFACTURE OF BRICKS.

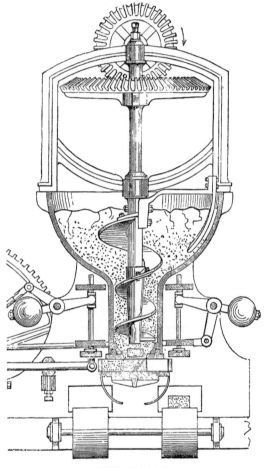

WET PROCESS.

The machine combines the three processes of crushing, pugging, and brick-making. The clay is dug and

spread over a large surface of ground, and tempered
with water by manual labour into a mass of soft clay,
by turning it over several times, and is then thrown
into a hopper, below which work a pair of crushing
rollers. There is an instrument with a number of
knives, or blades, to cram in the clay. The crushed
clay falls directly from the revolving rollers into a pug-
mill, the object of which is to incorporate the various
component parts of clay, and to make it of one uniform
character. The pugged clay is then forced out through
an aperture into a horizontal clay-box, or chamber,
where, as in the steam-engine by the action of a piston,
the clay issues in a stream on the delivery-table, where
it is cut into the desired thickness of bricks.

DRY PROCESS.

The dry brick-making machine works on the principle of forming the clay into moulds, instead of forcing
it through orifices or dies. The clay, as dug up, is
delivered to the machine ; or, if the material consists of
marl or fire-clay, or stony earth, it is crushed through
rollers, and delivered to the machine, which works it
into a dense mass with great solidity, and fills the
moulds of the form and size of brick required ; these
are arranged on a circular revolving table. Two moulds
receive their charge of clay at a time ; and while this
operation is going on, two other moulds that had been
previously filled are subjected to considerable pressure,
by pistons working on the opposite side. The finished
bricks as discharged from the moulds are delivered on
to a revolving endless creeper band for removal to the
kiln.

HOFFMANN'S KILN FOR BRICK AND LIME-BURNING.

The kiln is about 160 feet in diameter for a delivery
of 25,000 bricks per day.

The kiln is built in the form of a large arched passage, like a railway tunnel, bending round in going forward on the ground till it closes with itself to form a great circular ring-chamber, within which the burning of the bricks is carried on. Round its circumference there are twenty-four entrance door-ways, admitting of being closed, so as to retain the heat and exclude all entrance of air by the door-ways so built up. The great chamber may consist of twenty-four compartments or spaces, with one of these door-ways to each. In the centre of the ring a high chimney is erected, and from each of the twenty-four compartments of the annular chamber an underground flue leads into the chimney. There are, then, twenty-four of these flues converging towards the centre like the spokes of a wheel; and each flue has a valve, by which its communication with the chimney can be cut off. Arrangements are made by which a partition like a damper can be let down at pleasure, so as to cut off all communication between any of the twenty-four compartments of the ring-kiln and the next one. After being once kindled, the fire is never extinguished, but the burning of new bricks and the removal of the finished produce is carried on by a continuous and regular process from day to day. From one of the two open compartments men take out the finished and cooled bricks, and in the other one they build up newly-formed unburnt bricks. The air entering by these two compartments passes first among bricks almost cold, takes up their heat, then goes forward to warmer bricks, then to hotter and hotter, always carrying the heat of the cooling bricks forward with it till it reaches the part of the ring diametrically opposite to the two open and cold compartments. At this place it gets a final accession of heat from the burning of a very small quantity of of coal, which is dropped in among the bricks from time to time by numerous small openings furnished with air-tight movable lids.

ESTIMATES FOR A BRICK AND TILE WORK.

1. *Machinery for Wet Process, as shown on* PLAN P.

	£
Double clay hoist, self-acting, rails, complete, with 6 trucks with wrought axles, wheels, &c.	228
2 Improved brick-making machines, combining the three processes of crushing, pugging, and brick-making, each capable of producing 12,000 bricks per day, including dies and extras, applicable also for making tiles, drain pipes, &c. . . .	525
Steam-engine, horizontal, high-pressure, 20 H. P., complete, with boiler, hot-water heating apparatus, shafting, gearing, driving belts, and all iron work for Hoffman's kiln, complete . .	760
	£1,513

2. *Machinery for Dry Process, as shown on* PLAN P.

	£
Clay hauling apparatus, self-acting, with chain and rails, waggons, complete	130
Clay grinding mill, with fan, edge rollers, fitted complete . .	258
Elevators, self-acting, to carry clay from the grinding mill over brick machine, complete	45
Brick-making machine fitted up complete, with revolving table, moulds, &c., for making 14,000 bricks per day . . .	570
Brick press, to be fixed in front of brick machine, complete, with duplicate die, with creeper for carrying away the bricks, &c.	50
Steam-engine, 25 H. P., complete, with boiler, shafting, gearing, columns, water cistern, pump, drums, straps, and iron work for kiln, &c.	750
	£1,803

3. *Portable Brick-Making Machines.*

	£
Crushing, pugging, and brick-making machine, with dies and wire cutting, producing about 100,000 bricks per week, by *wet process*, complete, with steam-engine, boiler, &c. . .	650
Second-size machine, similar to the above, for producing about 75,000 bricks per week, with engine, boiler, &c., complete .	380
Third-size machine, without crusher, for clay free from stones, producing 9,000 bricks per day, arranged for working by animal power	155

£

Machine for tiles and pipes, also for solid or hollow bricks, worked by animal power 55

Tile pressing machines, from £25 to . . . . . ⎫
Mortar and loam mills, from £40 to . . . . . ⎬ 135
                                                      ⎭

Machine with pug mill, revolving table, moulds, steam-engine, boilers, with all fittings complete, capable of making 10,000 bricks per day by *dry process* 575

Clay crushing rollers, 20 to 25 in. diam., complete, from £25 to . 130

Combined rollers with pug mill, from £100 to 175

DIMENSIONS OF BRICK AND TILE WORKS.

Machinery as shown on PLANS P.

Scale: 16 feet = 1 inch.

	Length. feet.	Breadth. feet.	Square feet.	Height. feet.	Cubic feet.
Wet process machinery for making bricks . . .	46	38	1,748	15	26,220
Dry process machinery for making bricks . . .	70	27	1,890	20	37,800
Boiler house	27	9	243	8	1,944
Total of square and cubic feet			2,133		39,744

PLAN OF BRICKWORKS (WET PROCESS).

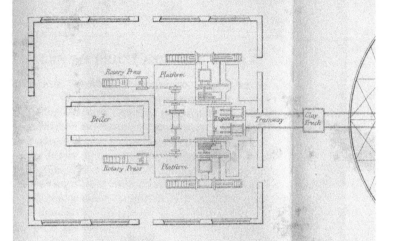

ELEVATION.

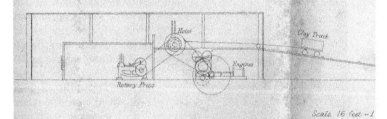

Scale 16 feet = 1

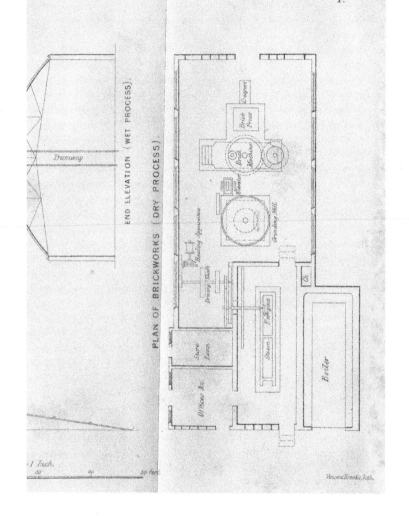

END ELEVATION (WET PROCESS).

PLAN OF BRICKWORKS (DRY PROCESS).

P.

REMARKS ON BRICK AND TILE WORKS.

BRICK-MAKING machinery should be adapted to the character of the clay. Some clays require for a good brick-making machine little or no preparation, while other clays require expensive machinery. Choice of site and selection of clay are therefore important preliminaries before ordering brick-making machinery, as the perfection and economy of the operations on the factory system will mainly depend on the quality of clay, and the general arrangements adopted in conformity therewith. Want of attention to this important point has been the chief cause of failure in the case of the Bombay Brick Works.

It is necessary that about half a ton of clay should be tested either in India or in England, to determine which class of brick-making machinery will answer the purpose, as, if in ignorance of the kind of clay to be worked, no sound opinion could be given.

Bricks being a very heavy article, the cost of carriage is also an important consideration, otherwise it will absorb all the profits. Brick machinery should be erected as near as possible to a town to save the distance of cartage.

From the estimates it will be seen that the machinery for making bricks by the dry process will cost more than by the wet process. But the great drawback to the latter is that the bricks are required to be cut by manual labour, and dried, and cannot be removed direct to the kiln as is done in the dry process. It is believed that the dry process of making bricks will be applicable in more cases, and though it requires more motive power to drive, it has certain advantages over the wet process. The latter, however, will answer very well if the clay is suitable for working by the machinery.

When ordering brick machinery, it is necessary to give the length, width, and thickness of the bricks required to be made. Bricks shrink in drying and burning about one-eighth of an inch per inch, for which allowance should be made.

Bricks of any size or form, and also roofing and paving tiles, are made by the same brick-making machine, by adopting suitable dies, moulds, and accessories.

Ornamental bricks are also made of a variety of colours, by mixing different substances with clay, which, by a judicious combination in the arches and other parts of a building, produce a very pleasing effect.

Estimates and plans have been given for both processes, making bricks either by the dry or wet process, and for permanent works for producing millions of bricks on the factory system. These include a self-acting hoist for discharge of clay, and other improvements. Brick-making machines are made of all sizes, adapted for temporary works and limited production, of which a price list has been also given. The total cost of the brick and tile works, as per estimates and plans, including buildings, kiln, &c., may be calculated at from £3,500 to £5,000. The cost of manufacturing bricks, English size, $9 \times 4\frac{1}{2} \times 2\frac{3}{4}$, when burnt, may be reckoned at about £1 10s. per 1,000 finished bricks;

this will of course vary more or less according to the price of coals, &c.

In manufacturing bricks the construction of the kiln for burning the bricks is a matter of very great importance. A description of the most recent improved kiln is given, and though it is much more costly in construction than the ordinary kilns, it has been stated to effect great saving in fuel.   The bricks burnt by this improved process are sounder and of better quality than from ordinary kilns.

STEAM MOTIVE POWER.

INTRODUCTION.

No one can doubt the great superiority of steam as a motive power for driving machinery, compared with 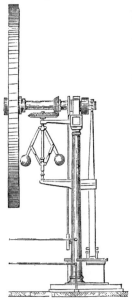 animal or even water power. The operation of the steam-engine is continuous, regular, and constant, and can be adjusted to the greatest nicety; whereas animal or water power is very fluctuating and irregular, and that unavoidably so. There are few places in India where the supply of water is abundant and constant throughout the whole year, and can be relied on as a means for driving machinery; hence, as a general rule, the use of the steam-engine is preferable, as a means of manufacturing with economy and certainty. The application of steam to all branches of industry is almost universal, and its importance and inestimable value are recognised throughout the civilised world. If only water power was to be made use of in India and other parts of Asia for driving machinery, the material prosperity to be derived from manufactures would be limited.

It was by the genius of James Watt that the power of the steam-engine was first applied in manufactures, in the year 1769; and since then it has been improved from time to time by eminent men in all parts of Europe, and by numerous modifications applied to all purposes of manufacture, driving machinery, impelling ships and railway trains, grinding corn, printing books, hammering, planing, and turning iron, and performing any description of mechanical labour where power is required.

As an economiser of labour and time the steam-engine has contributed more to human progress and comfort than any other mechanical invention; and as a motive power in the industrial arts it takes the precedence of all other powers.

A steam-engine consists essentially of a vessel, made cylindrical for giving the greatest strength, and fitted with a movable piston. The cylinder is placed either vertical or horizontal. All the other parts of the steam-engine are for the purpose of regulating the admission of steam, and for producing a rotary motion at the point where the power is required to be applied for working machinery.

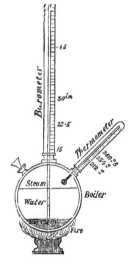

In connection with the steam-engine, the nature and properties of water are of great importance. Water is solid when the thermometer stands at 32° F., fluid at ordinary temperatures, and vapour or steam when the temperature is raised to 212°. But if this high temperature be not kept up the vapour is converted into water again. This property of reconverting from the form of vapour to that of water is invaluable in the steam-engine, by condensing the steam to its original bulk of water, and withdrawing that water and forming a *vacuum*. Another property of steam is expansibility or increase in bulk, the weight remaining the same. A cubic inch of water will give steam that will occupy 1,728 cubic inches, or one cubic foot of space. Steam also possesses the property of elasticity, as it has a tendency to free itself from the space in which it is confined. When the natural bulk of steam is compressed into half the space, the pressure is doubled, and so on. By cutting off the entrance of steam in any vessel or cylinder in which a piston moves, equal to half the stroke or length of the cylinder, the piston having received a momentum is propelled through by the expansive force of steam, and a quantity of steam and fuel is saved in proportion.

The elastic force of steam is regulated by the temperature under which it is raised, and increases in a much faster ratio than the temperature at which steam is generated. Thus at 212° steam exerts 15 pounds pressure to the square inch ; if the temperature is increased to 250°, the elastic force is double, or 30 pounds pressure to the inch ; and when the temperature is 400°, the force is 240 pounds to the square inch. The advantage, therefore, of using high-pressure steam is apparent. A small accession of heat at a high temperature produces an increase of elastic force, and the saving of fuel is in proportion to the increase of pressure. There is an invariable correspondence between the force of steam and the temperature at which it is raised.

To raise steam to a certain elasticity a certain quantity of fuel is required. If that steam be allowed, after having moved the piston of the steam-engine, to escape without having acted *expansively*, a portion of the fuel which was consumed to raise the steam up to that point of elasticity will be lost. The result of working steam *expansively* in the steam-engine is that power is made available by using the steam up to its last impelling force, and not allowing it to escape until the whole of that available force has been expended. If the steam from the boiler be cut off early when entering the cylinder, and expanded to its full available limits, then the highest pressure will have been used on the piston down to the lowest point. If steam be allowed to escape from the cylinder before its full force is expended to the lowest available pressure, the loss will be in proportion to the amount of the pressure not made available.

In the *condensing* steam-engine, when the steam is no longer required for working the piston in the cylinder, it is conducted by a pipe to another vessel, called the *condenser*, and, by the application of jets of *cold* water, is condensed into water. The water formed by the condensed steam is removed from the condenser by a pump ; a vacuum is thus produced, which assists the escape of steam from the cylinder. The engines constructed for working the steam on this principle are called *condensing* engines.

In *non-condensing* engines, the steam, when no longer required for working the piston, is allowed by a pipe to escape into the open air, and has to force itself against the pressure of the atmosphere. But there is no resistance when the steam is conducted into a condenser.

This is the essential difference between condensing and non-condensing engines.

The form of the steam-engine is either vertical, beam, or horizontal. The beam engine is heavy and costly, requires costly foundation, and occupies also a great deal of room. In the International Exhibition of 1862 two steam-engines were shown of 30 horse-power, by the same maker, both having the same power; the weight in the beam engine was 15 tons more than in the other. The horizontal form of steam-engine is now superseding the beam engine, on account of the simplicity of its construction and working, the less liability to break-downs, and the saving in first cost. A beam engine, say of only 12 horse-power, if it costs £320, the horizontal engine of the same power would cost £280. The advantages claimed for the beam engine are that the cylinder and piston wear better ; but it has been advanced that the cylinder made of a good metal is a long time in wearing, whatever may be its position; that the advantages claimed are merely supposed, and that the points of objection against the horizontal are not real. Moreover the form of the beam engine prevents the adoption of that speed of piston at which, all things taken into account, steam-engines work most economically.

The term *horse-power* was introduced by James Watt, to enable him to determine what size of steam-engines should be sent to his customers. He required them to tell him how many horses they were accustomed to employ to do the required work. It was found by experiment that the average force exerted by the strongest horses was equal to 33,000 lbs. raised one foot high in one minute, and this was taken as equivalent to one

horse-power. Since the time of Watt, however, the
capacity of the cylinder of the steam-engine answerable
to a horse-power has been increased, so that at the present time a horse-power is a term for expressing a
certain size of cylinder without reference to the power
actually exerted by the steam-engine.

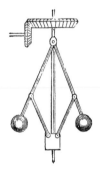

STEAM-ENGINES, BOILERS, ETC.

Condensing Engines are also called *low-pressure engines*. The steam in this class of engines, after having worked one way in the piston, is condensed by injection of cold water thrown by a feed-pump into the condenser. The water of condensation is drawn off by an air-pump; a vacuum is formed, the extent of which is seen at once by the rising or falling of mercury in a glass-tube fixed in the engine-house. Thus steam in a condensing engine is made to answer a double purpose, and the property of steam of reconverting into water is practically made use of. The cooler the water for condensing steam, the better is the result in working the engine; for which reason large reservoirs are required.

Compound Engines are high and low pressure combined. The principal feature in these engines is that the steam, after having been more or less expanded in a small cylinder, is further worked in a large cylinder; the expansion of steam is carried further for the purpose of ensuring economy of fuel; then the steam is condensed by jets of cold water, and the water withdrawn from the condenser by a pump as in the condensing engine. The condenser of a steam-engine for working in a hot climate is made of a larger size.

A High-pressure Engine is more compact than a condensing or compound engine, as no air-pump, condenser, &c., are required, and therefore it is also cheaper. The steam in the high-pressure engine is admitted from the boiler into the cylinder, as usual in the steam chest, where a valve works steam-tight. After working in the cylinder of a high-pressure engine, steam is not

condensed as in the condensing engine, and no vacuum
is produced; but it escapes into the air.   One side of
the piston of a high-pressure engine is open to the
atmosphere, the resistance of which it has thus to meet,

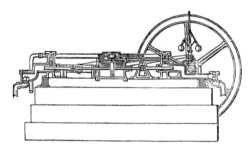

and the corresponding effect is produced on the piston;
this is not the case in a condensing or compound
engine.

The Cornish or Cylindrical Boiler is the prevailing
type of stationary boilers in England, and is now work-

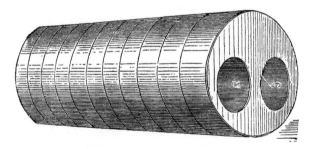

ing in many mills and other establishments in India.
It consists of a cylindrical shell with one or two flues
passing through it from end to end, of which the near
ends are fitted up as furnaces.   Some of these boilers
are constructed with a number of stays, uniting the top
and bottom, so that a free circulation of water is main-

tained, and a free delivery of steam effected, while a
considerable addition to the heating surface is ensured.

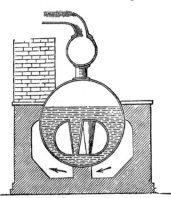

This type of boiler is now generally adopted for its
general ability to fulfil the various requirements in
practice.

The Fuel Economiser, adopted for utilising the waste
heat from boilers, consists of a series of pipes, placed in
the main flue behind the boilers, through which the
smoke passes on its way to the chimney. The patent
consists in the use of self-acting scrapers that work on
centres on the pipes, furnished with a steeled thin
cutting edge, by which the soot deposited on the pipes
by the smoke is effectually removed by the creeping
continuous motion given to the scrapers, without any
attention on the part of the fireman. The boilers can
be worked whenever desired, independently of the
economiser, by a reserve flue and damper.

The *Indicator* is a piece of mechanism by means of
which, at any time when the engine is in motion, the
working condition of the engine, the force of steam, and
the quantity of coal consumed per hour, are known. A
diagram is traced on a paper by the revolving cylinder
by a pencil, and by means of a scale the distance on the

diagram is measured, and from that the exact pressure of steam per square inch upon the piston of the engine. As a satisfactory method of knowing the working capabilities of a steam-engine, this little instrument is now

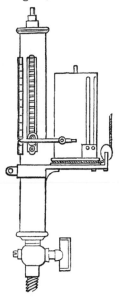

in all well-conducted factories in every-day use ; seeing that whenever any part of the engine gets out of order and is working defectively, the fact is known at once.

The *Injector* is now extensively employed as a substitute for the ordinary feed-pumps, where a boiler is used and steam produced. It is entirely independent of the engine, and is put in operation at once by simply opening connections with the boiler. As it acts when the engine is at rest, it supersedes the donkey-engine, and is very useful for filling the boiler with water when the engine is not working.

PRICES OF HORIZONTAL HIGH-PRESSURE STEAM-ENGINES AND BOILERS, WITH ALL IMPROVEMENTS.

Horse Power.	Diameter of Cylinder.	Length of Stroke.	Number of Boilers.	Length and Diam. of Cornish Boilers.		Weight of Engines, Boiler, &c.		Price.
	Inches.	Inches.		Length.	Diam.	Tons.	Cwt.	£
4	7	14	1	8	$3\frac{1}{2}$	4	15	96
6	9	18	1	10	4	5	5	144
8	10	20	1	14	4	6	15	192
10	11	22	1	13	5	7	16	240
12	12	24	1	15	5	9	5	288
14	13	24	1	$16\frac{1}{2}$	5	10	10	336
16	14	24	1	18	5	11	10	384
18	15	30	1*	18	$5\frac{1}{2}$	12	10	432
20	16	30	1	20	$5\frac{1}{2}$	13	10	480
22	$16\frac{3}{4}$	30	1	20	$6\frac{1}{2}$	14	15	520
24	$17\frac{1}{4}$	36	1	$21\frac{1}{2}$	$6\frac{1}{2}$	16	10	576
26	18	36	1	22	7	18	5	624
28	$18\frac{1}{2}$	36	1	$23\frac{1}{2}$	7	20	0	672
30	$19\frac{1}{2}$	36	1	25	7	22	0	720
35	20	38	2	22	$5\frac{1}{2}$	25	10	840
40	24	40	2	25	$5\frac{1}{2}$	27	15	980
45	26	48	2	24	6	34	0	1075
50	28	56	2	28	6	39	0	1120

* Instead of one, two boilers of smaller diameter will be preferable.

Prices of Condensing Engines may be calculated about £6 additional per horse-power.

PRICES OF PATENT FUEL ECONOMISER.

Boiler.		Pipes required.	Weight.	Price.	With casing.*
Length.	Diam.		cwt.	£	£
13	4	16	40	60	72
14	5	18	45	75	90
15	5	24	60	90	108
18	5	32	80	120	144
18	5½	32	80	120	144
20	5½	40	90	151	180
20	6½	40	90	151	180
21½	6½	40	90	151	180
22	7	48	120	181	217
25	7	48	120	181	217

* A brick wall round the economiser answers all the purpose of iron casing, but the advantage of the latter is that it can be removed and refixed whenever wanted for repairs, &c.

PRICES OF PORTABLE STEAM-ENGINES MOUNTED ON BOILERS ON WHEELS, COMPLETE FOR WORKING.

Power of Engines.	Cylinder.	Price.	Power of Engines.	Cylinder.	Price.
		£			£
4 Horse.	Single.	140	14 Horse.	Double.	390
6 ,,	,,	175	16 ,,	,,	415
8 ,,	,,	210	18 ,,	,,	465
10 ,,	Double.	270	20 ,,	,,	505
12 ,,	,,	300			

REMARKS ON STEAM-ENGINES, ETC.

THE primary object of steam-engines and boilers intended for working in India and the East, is that they
should be designed and constructed specially for economy
of fuel; as fuel, generally speaking, is more costly than
in England. The cost of the engines themselves,
therefore, ought to be a secondary consideration. The
price list of engines is for the best, with all the recent
improvements. The class of steam-engines suitable for
India and other places in the East is no doubt the condensing engine, or the compound engine, or high and
low pressure combined, provided the supply of water
for condensing steam is quite ample throughout the
year; otherwise, for want of sufficient water, great
difficulty will be experienced in working these engines,
and even at times entire stoppage of the works will be
the consequence. As some provision against any un-
foreseen shortness in the supply of water, it would be
well to have an arrangement in the condensing engine
to throw the exhaust steam into the atmosphere, and
shut the passage into the condenser. In some situations
the condensed water from the condensing engine may
be usefully employed for driving smaller machinery.
The consumption of fuel in a condensing engine will
be about 15 per cent. lower than in a high-pressure
engine. Non-condensing or high-pressure engines are

cheaper than condensing or compound engines, as there is no condenser or air-pump attached. Where the supply of water is not abundant these engines will suit well, and will work without trouble ; but the consumption of fuel will be more than in condensing or compound engines. The principle of working steam *expansively* has been applied to both classes of engines. Various forms of valves are used in the construction of steam-engines for working the steam economically, and recently Corliss's valve gear has been applied, in some of the largest textile factories in England, with the advantage of saving more than ten per cent. of fuel. Not half the number of engines sent to India for the cotton mills, even by the best makers, had the cylinders clothed for economy of steam ; but on account of more competition, the makers have commenced to design engines to ensure economy of fuel, about which they were at first indifferent. Some of them have made it a rule to clothe the cylinders of each steam-engine made by them with hair felt, mahogany, and iron casings, and a steam jacket to maintain a suitable temperature in working steam expansively. The condenser is also made of a larger size than usual, for working better in a hot climate.

The class of engines employed in almost all the mills in Bombay is the *horizontal compound engine,* high and low pressure combined. An example may be cited here of the most approved form. If the diameter of the high-pressure horizontal cylinder be 18½ inches, that of low-pressure or condensing cylinder will be 30 inches, both with 5 feet stroke ; the vertical air-pump will be 22 inches diameter, 2 feet stroke, condenser 26 inches diameter and 50 inches high, and cylinders fitted with valves and apparatus for cutting off at different portions of the stroke ; with plates for bottom of engine, foundations, and for carrying air-pump and condenser.

The class of boilers that will answer well in India, China, and other places, is the boiler with patent taper stays, as it generates more steam with less fuel than any other boiler in ordinary use. The generating power of this boiler has been stated to be that one pound of ordinary coal will evaporate $9\frac{1}{4}$ to $9\frac{1}{2}$ lbs or

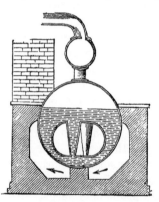

nearly one gallon of water per hour ; while a common two-flued boiler without patent stays evaporates $8\frac{1}{4}$ lbs ; and in a double boiler, mounted one over the other, sent to Bombay, it was only $6\frac{1}{2}$ lbs with Welsh coal. Boilers should be above the ground level, otherwise it has been found from practical experience that in the monsoon, when the ground is very damp, there is great loss of time and fuel in getting and keeping up steam.

In 1866 a boiler was patented in England and worked in the Paris Exhibition, composed of wrought-iron tubes. The advantages claimed for this boiler are, that there is great economy in getting up steam; the circulation of the water is well kept up within every part, as each tube has within it an internal one rising up to the water space; no joint or bolt is exposed to the action of the fire; the tubes can be examined and replaced in a quarter of an hour; it can be packed up in a small compass for exportation; and the bursting pressure of each tube being 2,000 pounds per square inch, there is no risk of explosion.

The *Fuel Economiser*, for which the price list is given, is one of the important practical inventions for saving fuel, by using up a considerable amount of surplus heat, which otherwise would escape up the chimney. The economiser is very simple in construction, is not very expensive, and works satisfactorily. Even in England, where coal is cheap, it has been applied in many factories, and the proprietors have testified in writing, after working it for some years, that it effects a saving of fuel from ten to fifteen per cent. It is now working in the Bombay cotton mills recently erected. It is made of all sizes, and is applicable to boilers of any description.

For consuming smoke from chimneys, which is a source of nuisance, numerous patents have been taken out. One of the expedients adopted in Lancashire is this: A supply of air for consuming the smoke is introduced at the front of the furnace, and adjusted to the wants of the furnace by jets of steam. The steam nozzles with the air tubes possess the power of urging and creating a draught, with such precision as to sweep the whole surface of the fuel. But even with all improved appliances, an attentive stoker is absolutely

T

necessary, as much depends on the way the boiler is
fed with coals, there being a right and a wrong way
in this as in other matters. If care be not taken to

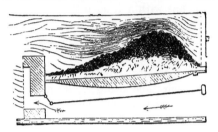

give the fuel enough air, it will smoke. The preven-
tion of smoke depends upon perfect combustion, and

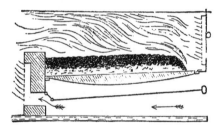

this leads to economy in fuel. A self-acting feeding
furnace to work with economy has not yet been contrived. Portable engines, and engines with their boilers
on wheels, are now extensively used in England where
formerly horse-power was employed, and are of special
utility for supplying steam power for agricultural operations, and temporary works. Directly on the top of the
boiler a high-pressure steam-engine with a single or
double cylinder is fixed. Portable double *expansive*
engines are now made, of 6, 8, 10, 12, or 14 horse-

power, so improved that the makers guarantee that the consumption of fuel will not exceed respectively 2, 3, 4, 5 and 6 cwt. of coal per day.

The nominal horse-power of a steam-engine does not express the actual power. Steam-engines are worked and proportioned to exert three times, and even more than three times, the nominal power. Hence the proportion between the actual power exerted and the nominal horse-power is varying, the dimensions of the engine expressed in nominal horse-power not exactly expressing the power that will be exerted.

By burning four pounds of fuel, more or less, according to the construction of the engine, boilers, and management, a mechanical effect of 1 horse-power is obtained, and the standard of 1 horse-power applied to the steam-engine is 33,000 pounds raised one foot high. This standard was fixed in a court of law in England, in which the question arose as to what constituted a nominal horse-power in a steam-engine.

It has been shown, from actual practice, that the *indicator* is as useful to a person in charge of a steam-engine as a compass is to a navigator. By testing the engines with the indicator, the defects are at once laid bare, the valves are properly set, and a great saving of fuel is the result, besides additional power obtained. This has been demonstrated and proved in several cases. In one mill, after the engines were indicated and the valves properly set in, and the steam cut off at the proper length of stroke, the consumption of coal was reduced by nearly half. In another mill the consumption of coal was $7\frac{1}{2}$ lbs. per hour per horse-power. The engines were indicated, but after the proper setting of the valves, increasing the pressure of steam, and casing the cylinders, the consumption of coal was reduced to 5¼ lbs., besides the advantage of the addition to power gained. These examples show the necessity of indicating the engines every day ; but though the indicator does not cost more than a few pounds, with rare exceptions it is not made use of in the Bombay cotton mills, more on account of the incompetency of the persons in charge of it than any other cause, proving that cheap incompetent engineers are very dear in the end.

It has been stated that the consumption of coals in the Corliss engine, which is now made by more than one firm in England, is 2½ lbs. per indicated horse-power per hour.

Gas engines are now made in England, and are used in certain situations in centres of towns, where a very limited power is required, as in small printing establishments or very small workshops. The gas engine resembles in general appearance the horizontal steam-engine, but requires no steam, coal, or chimney. The gas with which Bombay and Calcutta are now lighted is the agent which imparts a regular motion in the engine. The engine consumes about 90 cubic feet of gas per horse-power when working, is free from danger, and is started or stopped by merely turning a gas cock. In a waterfall the power generated with any given quantity of water is measurable by the difference of level between the highest and lowest points.

STEAM-ENGINES AT THE PARIS EXHIBITION OF 1867.

(From the *Engineering Journal,* May, 1867.)

There is an American steam-engine which arrests the attention of every competent engineer who walks round the circle of that department. It embodies all the characteristics of American engineering ; it is light in structure, elegant in design, refined in artifice, charming in its simplicity, and yet there is nowhere any sacrifice of perfection or usefulness for a fictitious semblance of simplicity. Where anything is to be done, it is done straightforwardly and to the purpose ; there is no pretence of doing two things at once, while seeming to do one. The engine is called the Corliss engine ; it is not a novelty in America, and is not unknown to well-informed English engineers. Nevertheless, we may safely say it is a practical novelty, and the manner in which it is designed and executed is admirable. Its special peculiarities are its valves, and the mode of working them, and its general disposition and structure. It is made like all American engines, with saving of fuel as the great consideration. Because we have cheap fuel, we have too often to pay the penalty of bad engineering. The Americans make fuel out of brains, and a Yankee engine doing the same work as one of ours on half the fuel at once deprives us as a nation of any manufacturing advantage from our cheap and abundant fuel. The Corliss engine spares every ounce of coal ; it is always on the watch to do so, and so lively is it that the moment the engine goes its speed, and can spare a little steam or coal, the watchful mechanical governor of the engine takes to sparing the steam and hoarding the fuel. Just as a good locomotive driver is always watching for the opportunity of putting the expansion gear on a new notch to increase his cut-off, so is the mechanical government of the Corliss engine carried on by mechanism which successfully represents the care and skill of an accomplished driver. The engine has four ports at what we may call the four corners of the cylinder. They are worked by four independent valves. The object of this is to spare all passages and all waste in them. The engine has detached the escape-valves from the steam-valves, so that both may have independent motions ; but to the steam-valves is added what we may call a detaching movement, and this "detent" movement is one of the characteristics of the engine. It is indeed very peculiar, original, and American ; it is the means by which the mechanical governor, ever on the watch, renders the engine "autonomic." It acts thus : the valve-rod of each steam-port is held, as it were, between the finger and thumb of what one may call the right arm and the left arm of the governor. These two arms are moved regularly back and forward by an unvarying motion ; but the finger and thumb of each hand of the governor let go their hold of the steam-valves the moment the governor judges that enough steam has entered; and the moment the opened finger and thumb release the valve, no more steam enters, and the rest is saved. When a heavy load is suddenly put upon the engine, the finger and thumb hold on longer, and let more steam enter. The moment the load is lightened and the engine can go easy, the finger and thumb let go quickly again, and so spare fuel.

We must candidly say that this arrangement is fascinating and elegant. It would be easy to do it clumsily and badly, for letting a valve go with a sudden jerk is neither good engineering nor sound economy; but Mr. Corliss has two delicate little air-cushions placed immediately under the drops where the finger and thumb release their hold, and so the whole of these motions, however quick and sharp, are gentle and soft. In regard to the general structure of the engine, it is almost superfluous to say that forethought, judgment, and skill are equally displayed in the wise disposition and good proportions of its parts. It is an engine from which any good English engineer may read off to himself an agreeable and interesting lesson. And, indeed, more engineers than one have already read that lesson, and in our own English department may be seen two engines, which, if not copies or literal imitations, are plainly its progeny, sprung from its seed. Hick, of Bolton, and Whitworth, of Manchester, have plainly adopted the opinions to which we have just given expression, and they must long ere now have carefully studied this engine, and bethought themselves how to use or better the example; and of these two, one at least is certainly on the way to better the example. That the firm of Messrs. Hick should have appreciated the merits and introduced the improvements of such an engine into English use is highly creditable to their judgment, and is just what might be expected from a firm which is notorious for having very early introduced, and very energetically maintained, the Woolf system of double-cylinder engine in this country. The engine they have made is a good and economical one, and if they have not adopted the latest refinements of the American engine, they have perhaps gone as far as the ordinary demand for refined engines would allow.

The engine, however, in the English department which is most deserving our attention is one which steps far beyond any other steam-engine in the Exhibition in its character and purpose. We do not say that it has achieved all its aims; but it has gone so far, and accomplished so much, that we are bound to regard it as an engine of the highest promise: curiously, too, it is a hybrid or cross-breed of engine; it is derived from Corliss's; it is improved by two Americans, Allen and Porter, and it is constructed with the forethought, proportion, symmetry, and truth of construction which have so long distinguished all that issues from the establishment of Whitworth, of Manchester. The design of this engine, as far as can be gathered from its structure, is this: it is meant to be Corliss's engine, but without Corliss's valves and without the finger and thumb motion which distinguishes it, but it dispenses with these valves and these motions only by the substitution of something else, which is at once the same and different. Its valves are slide valves, one an ordinary slide and one a cut-off slide; but the character of the slides is very peculiar, and the mode in which they are moved, although not Corliss's finger and thumb motion, are meant to obtain from the gentle movement of the ordinary eccentric an action of the valves equivalent to that of Corliss's by a gentle and slow, instead of a sudden and quick, movement. The diagrams of the engine show that the valves perfectly succeed in producing in the cylinder a high and economical development of steam power. So far, however, we can say little, except that they have constructed a Corliss engine, or its equivalent, in another manner.

Having overtaken Mr. Corliss in the application of the steam in his cylinder, the constructors of this engine now make a great stride to go beyond him. They say, " Our valves are not only as good as Corliss's, but they are, withal, so smooth and gentle in their action, that they are capable of working much more rapidly." They, have, therefore determined to use this smoothness of action for the development of a far higher amount of power out of an engine of given size than had heretofore been accomplished. Corliss's engine, like other fixed steam-engines, performs admirably with a piston travelling, say, 200 feet a minute, or about the usual traditional speed of the steam-engine. The Allen-Porter-Whitworth engine leaves this behind with a long stride ; its piston starts away at the unusual speed of 800 feet per minute, and, in doing so, quadruples the work done by an engine of given size and power. This is certainly an unparalleled feat in the gymnastics of the steam-engine, and, if successfully accomplished, seems to promise an important revolution in machinery.

It is not to be overlooked that we are speaking of a condensing-engine, and, what is more surprising still, an engine whose air-pump works with the same speed as its steam-piston. The eye scarcely can see the plunger of the air-pump clearly, from the rapidity with which it travels in and out of the condenser. The plunger looks more like the elongated shot of a Whitworth cannon than the piston of an air-pump ; in shape it is truly an elongated steel or iron shot, which strikes the water in the air-pump with such velocity that if the point of the plunger were not shapened into a parabolic curve its stroke on the water would shatter the condenser to pieces. As it is constructed, however, by means of ingenious hydraulic mechanism, the rapid stroke of the air-pump is converted into so gentle a rise and fall of the water that the valves work with scarcely a sound, and a gentle throb when your hand is laid on the condenser is all that tells you of the pulsation going on within. The engine is a marvel of ingenuity and design. It is, however, neither unkind nor unjust to say that the engine is still on its trial, that trial which alone tells us whether any mechanical engine is destined to live and advance our generation, or to lose its reputation with its novelty, and die out.

COAL MINES IN INDIA AND THE EAST.

INTRODUCTION.

COAL is superior to every other description of fuel, and bears a most mportant relation to the establishment of mills and factories in India, China, and other parts of Asia. The extent of its distribution in the East will be seen from some valuable papers given on that subject from the highest authorities. The East India Railway, in Bengal, which extends more than 1,200 miles, passes near the mines of Burdwan, Raneegunge, and obtains its supply of coal at a cost of *ten shillings per ton* at the pit's mouth.

In the Bombay Presidency the Great Indian Peninsular Railway Company, with the sanction of Government, has made a contract with the Nerbudda Coal and Iron Company for a supply of coal at prices ranging from £1 4s. per ton, decreasing every year as the mines are developed, down to 18s. per ton in January, 1872. There is little doubt that with the progress of railways, roads, and canals, the coal, iron, and other mines will be worked extensively, and will be a source of profit to the people, as well as to the companies and to the Government.

The most rudimentary form of coal is *peat*. There is also another kind of coal of a brown colour known as *lignite*, showing a woody structure. But the variety of coal generally used in factories for steam purposes is *bituminous* coal. Another variety is *cannel* coal. When a single piece of this is ignited it continues to burn with a smoky flame; hence the original name of "candle coal," corrupted into cannel coal. It is particularly suitable for gas-making, and for this reason commands the highest price.

The amount of power contained in coal is almost incredible. In burning a single pound of it, there is force developed equivalent to that of 11,422,000 lbs. weight falling one foot, and the actual useful force got from each pound of coal in a good steam-engine is that of 1,000,000 lbs. falling through a foot; that is to say, there is *spring* enough in coal to raise a million times its own weight a foot high. In this mineral there is what all the world wants—power. Coal is the product of the *decay* of vegetable matter, of which fact evidence is to be found in the different samples of coal to be seen in the London Geological and other museums. Vegetable matter consists essentially of carbon about 50 per cent., water 47, and ash 3 per cent. Now, in proportion to the degree of decay will be the relative increase in the per-centage of carbon in coal, and the final stage is reached in the variety of coal termed *anthracite*, which may contain upwards of 90 per cent. of carbon. Between unchanged vegetable matter and anthracite every gradation is observed. An excellent illustration of this kind of change is presented by a peat bog, where moss, which in Europe is the source of peat, may be seen growing at the top and gradually passing into peat underneath; and at the bottom decay may have so far advanced as to yield black peat free from all appearance of vegetable structure. What takes place in this transformation of moss into peat is precisely similar in kind to what takes place in the conversion of vegetable matter into coal. The difference between the two cases is simply one of degree. Just in proportion as decay progresses in moss, will the proportion of carbon in the product relatively increase.

COAL MINES IN INDIA.

(From a Paper by M. C. BROOKS, ESQ., Indian Department, London, in Record of the Great Exhibition, 1862.)

The coal resources of India are fairly represented in the present exhibition. There is an excellent series of coals sent by Professor Oldham, Director of the Geological Survey of India, which may be accepted as types of the principal Indian coalfields. From the mines in the Singaran valley specimens are shown from Chokidanga, Toposi, Mangulpur, Harispur, and Babusol. The colliery at Chokidanga was established in 1834. The seam of coal is $15\frac{1}{2}$ feet in thickness, and the average annual yield of the last three years has been 360,000 maunds. At Toposi a seam of 22 feet is worked, and last year 300,000 maunds of coal were raised. Mangulpur is a long-established colliery, where a seam of $15\frac{1}{2}$ feet (including 9 inches of shale) is worked, and yields annually a million maunds of coal. Babusol and Madhubpur, or Harispur, are situated in the lower portian of the Singaran stream, and are the most eastern collieries of the field; they unitedly yielded 524,000 maunds of coal during last year. The Central Ranigunje coalfield is represented by specimens from Ranigunje, Rogonathchuk, Bhangaband, and Bansra. Mines were actually worked in the Ranigunje field as long since as 1777. This field is an important one, not only for its extent, but its position. It lies at a distance of from 120 to 160 miles N.W. of Calcutta. The area as at present ascertained is about 500 square miles. The most extensive workings are Ranigunje, near the Damuda river. The entire seam is 13 feet in thickness, divided by a band of shale into two seams of 9 and 3 feet. From each of these specimens are sent. The Ranigunje workings yielded 1,600,000 maunds in 1800-1. Rogonathchuk is on the banks of the Damuda, and is one of the oldest collieries in the field. The seam is $12\frac{1}{2}$ feet in thickness. Bhangaband is in the same neighbourhood, and yields annually about a quarter of a million maunds. At Bansra the seam is about 7 feet thick. Of the mines in the Nunia valley there is a deficiency of samples. Those in the eastern division, thirteen collieries in all, are unrepresented; those in the western are known only by a specimen from Futtipur, on the Grand Trunk road. The bed here is of ten feet thickness and excellent quality. From neighbouring mines specimens are shown from Hahinal in the west of the Ranigunje field, near the confluence of the Barakar and Damuda; from Chinakuri on the Damuda, not far from Hahinal; from Dumarkhunda to the west of the Barakar; and from Kastae in the extreme north of the Ranigunje coal-field. The out-turn at the colliery of Chinakuri is not less than three and a quarter millions of maunds annually. At Kastna there is a seam of upwards of 30 feet in thickness, which is worked in open quarries. The specimen sent is from the lower 11 feet of this seam. The produce of all these fields is not less than 320,000 tons per annum.

The coal in the lower Damuda is frequently found intersected with basaltic trap, and in most cases the structure of the coal is entirely changed. The coal becomes prismatic or columnar, and this may be seen over large areas. The columns are often not more than half an inch in diameter, and so completely separated that it is difficult to procure a specimen which will show more than one single prism.

The following table represents the relative value of the different coals of which specimens are shown from the coalfields of Bengal:—

Name.	Thickness of Seam in Feet.	Composition of Coal.		
		Carbon.	Volatile Matter.	Ash.
Kurhurbalee . . .	7 to 16	66·70	24·80	8·45
Futtehpur	10	63·80	25·00	11·20
Dumarkunda . . .	10	62·40	22·60	15·00
Kastna	30	61·40	28·00	10·60
Chokidanga	15½	56·80	34·00	9·20
Chinakuri	10½	53·20	35·50	11·30
Hattinal	11	52·60	33·00	14·40
Madubpore (Harispur).	17	51·10	35·40	13·50
Ranigunje	9	50·80	36·00	13·20
Do.	3	50·30	36·30	13·40
Toposi	22	49·20	35·40	15·40
Bansra	13	47·00	40·00	13·00
Rogonathchuk . . .	10½	46·90	35·00	18·10
Babusol	17	46·00	35·40	18·60
Chilgo	5	45·50	43·50	11·00
Oormoo	7 and 3	45·00	44·60	10·40
Panchlyni	7	44·20	34·10	21·70
Mangalpur	15½	43·90	38·40	17·70
Bankijora	19	43·50	42.00	14·50
Banali	12	42·60	44·20	13·20
Bhangaband . . .	7	40·30	28·40	31·30
Bhorah	17	25·20	37·20	37·60

Of the collieries in the Rajmahal hills, those in the
neighbourhood of Brahmini Nuddi are represented by a
specimen from Panchbynee, in the extreme south of the
hills; those near Banshi Nuddi by specimens from
Oormoo, Chilgo, and Bankijora. These are often
spoken of together as the Alubera collieries. At
Oormoo the seam is 10 feet, at Chilgo 5 feet, and at
Bankijora 19 feet in thickness. The field on the north
and north-west of the hills is represented by a specimen from Bhorah. There are also specimens from the
Khurhurbari coalfield, and from Bareilly. The districts of Palamow, Khasia hill, Singrowli, and Scinde are
unrepresented. The amount of coal raised in the
Singrowli coalfield is small; coal is known to exist in
a westward direction from this bed towards Singhpoor,
but no collieries are yet established. The Nerbudda

valley also contains coal, of which three specimens are sent from Mohpanee, in Nursingpore. The coal in this locality forms a strip or band of irregular width along the foot of the Puchmurree hills. There are three seams at Mohpanee of 10 feet, 6 feet, and 3 feet 6 inches respectively. These mines have been leased to the Nerbudda Coal and Iron Company. The whole amount of coal raised in India during the year 1866 was ten million of maunds, or about 370,000 tons.

Coal Mines in Assam.—A cursory examination of the coalfields of Assam has established the existence of coal mines both at Jaipore and Terap, the latter of which especially are said to promise an unlimited supply of very superior quality.

It has been determined at once to throw open the coalfields of Assam to private enterprise, and to grant leases on certain conditions; and the grant of a perpetual lease of five plots of land in the Sonthal pergunnahs which contain coal has been authorised on the same terms. In the Central Provinces coal has been found in eleven different places in the Pench valley, with a thickness of good coal in the seam varying from 1 ft. to 12¼ ft.; the angle of dip is everywhere very small, from 3° to 10°, and the quality and depth of most of the seams are supposed to be constant, at all events over considerable areas.

COAL MINES IN CHINA.

(From the *Journal of the Society of Arts.*)

Extensive mines of coal exist in the mountains to the north-west of Pekin. It costs about 16s. per ton at the pit's mouth, and more than double this amount per ton is paid for transport to the coast; but the mines are worked in the rudest way, and the little coal that finds its way from the western ranges to Tien-tsin is conveyed on mules or camels from the mountains to Tung-chow, on the Peiho, and thence down the river in boats to this port. From the mines in the northern range there is water communication of an indifferent character to Tien-tsin, but the quality of this coal is much inferior to that which comes from the western mountains. Here, however, is a great source of wealth, only waiting the application of European skill and capital to enrich those who undertake its development. There are three descriptions of native coal to be purchased in Hankow. One known in China as dry coal, is retailed at about 600 cash per picul; that known as smoke coal is quoted at about 750 cash per picul; and a third, which is called white coal, costs about 800 cash per picul. The dry coal is a sort of coke, and is admirably adapted for all household purposes. The smoke and white coal are well suited for and employed by steamers. Hankow is furnished with coal by the Hunan coalfields, the position of which can be determined by a glance at any ordinary map. By following from its source the river (the Hsiang-Kiang), which, rising in the Hsiao Ling mountains, flows northwards until it reaches the Tung-Ting lake, the district city of Kyang will be found situated a short distance above the point where the north-east corner of Kiangsi cuts into Hunan. Here are the mines which supply the yen-mei, or smoke coal. Proceeding north we reach the great mart of Hsiang Tan, situated at the junction of two branch streams with the Hsiang-Kiang. The more westerly of these streams flows past a city marked on the map as the district city of Syang-Syang, and it is in this neighbourhood that the ku-mei, or dry coal, is produced. Further north two larger tributaries, also from the west, swell the volume of the main river; thirty miles above this embouchure they unite, and at the point of union is the district city of Fyang, close to which are the hills which yield the pai-mei, or white coal, a description of anthracite. On the Yang-tze Captain Blakiston saw no coal until he was forty miles beyond Chang-Fu, that is to say, over 440 miles above

Hankow. According to the Chinese the coal produced in Sze-chuen and the western part of Hu Pei is inferior to that which comes from Hunan, a statement which would seem to be confirmed by the fact of Captain Blakiston seeing at Sha Skit, about 190 miles above the outlet of the Tung-Ting lake, junks laden with Hunan coal bound upwards. Many are of opinion that coal should be found much nearer to Hankow; and during the past two years two foreign firms—one British and one American—have, with the greatest perseverance, been endeavouring to trace its existence in the Ching-kow hills, ten miles above Hankow. The British firm has recently desisted, but the American firm still continues its researches. Many of the hills between Kin-kiang and Hankow present every appearance of being rich in immense treasures, more especially that fine range which terminates eighty-five miles below Hankow, in the picturesque bluff known as the Cock's Head.

(From the Quarterly Review, 1866.)

Pekin Coal.—We may state that having examined coal, including lignite, from nearly every part of the world, we have seen no steam-coal superior to that from the neighbourhood of Pekin, where it is reported that a magnificent coalfield exists not less than 300 miles in extent. This report is founded on a personal communication from a geologist who has spent three years in the exploration of that coalfield. We have received several samples of Chinese coal, and we find them to differ much in quality. In 1862 they were tried in some of Her Majesty's ships under the command of Admiral Sir James Hope, and the results obtained were precisely such as the composition of these coals, which has been accurately determined, would indicate. The day will arrive when the coal-mines of China will prove a source of wealth and power, and may possibly determine who shall exercise naval predominance in

the East. These mines, in order to their successful
development, must be worked under the direction of
colliery engineers of experience and skill, and will
require the introduction of steam winding and pumping machinery. But the Chinese authorities dread all
such foreign innovations, especially as they must for a
considerable period be under the supervision of despised
barbarians. Perhaps they may be disturbed by a
vision of what has befallen India. If, unhappily, we
should again be involved in war with the Celestials,
and again be victorious, it might be well to stipulate
for a concession of a portion, at least, of this great
coalfield.

Much interest has been excited in the public mind
concerning the recent discoveries of very large accumu-
lations of petroleum in Rangoon and in the United
States. This substance is a combustible mineral oil,
composed essentially of carbon and hydrogen, which
may be employed either as fuel or for the purpose of
illumination. Petroleum, or rock oil, has long been
known to occur in various parts of the world. It is
derived exclusively from vegetable or animal matter,
and in many cases has certainly been produced from
coal by a natural process of distillation. According to
Professor Lesley, of the United States, one of the best
authorities on the subject, rocks hold it in three ways—
by being more or less gravelly or porous throughout;
by being cracked in cleavage planes throughout; and
by being traversed by large fissures, which are probably
all of them mere enlargements of cracks along the
cleavage planes. We have only to observe that *quâ* fuel
we may *virtually* regard it in the same light as coal.
What is petroleum but the essence of coal, distilled
from it by terrestrial heat ?

QUANTITY OF COALS EXPORTED FROM GREAT BRITAIN.

To	1866.	1865.	1864.	1863.	1862.
	Tons.	Tons.	Tons.	Tons.	Tons.
India .	436,292	342,283	364,038	357,997	273,146
France .	1,904,091	1,589,707	1,447,494	1,306,255	1,443,115
Russia .	575,154	488,178	472,844	460,176	438,190
Denmark	696,781	545,333	598,282	560,864	568,879
Prussia .	476,529	597,771	355,722	522,300	535,336
Hanse Towns	611,315	604,760	570,590	215,202	254,314
Spain. .	527,181	473,301	546,029	567,872	523,258
Brazil .	245,321	222,985	186,992	166,085	138,543

FACTORY BUILDINGS.

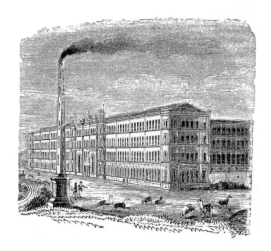

INTRODUCTION.

Situation.—Before erecting any buildings for factories or mills, great attention should be given to the question whether they should be in a town or at a distance; and also to proximity to rail, river, or canal. Pure and abundant water will be required for all factory purposes, but more in bleaching, dyeing, calico-printing, and paper-making, than in other works. The situation should be also as near as possible to the land producing the raw materials, to the market for manufactured goods, and to the supply of fuel. Labour which can be relied on at all seasons will also be necessary. Wherever these exist in combination, more or less, manufacturing processes can be carried on with advantage.

Size.—The advantages of manufacturing with improved machinery will only be derived by conducting it on a scale of sufficient magnitude. In that case, the division of labour is profitable, the raw material is economised, and the management and superintendence are confined within narrow limits. Up to a certain extent, the cost of production will be less in proportion to the magnitude of the works. This fact has been kept in view in fixing the size of works for which plans and estimates are given.

Plans of manufacturing establishments will vary according to situation and other local circumstances. In a majority of cases, however, the plans of factories given in this work will be applicable, or could be easily modified and altered to suit particular localities. The plans, though necessarily on a small scale, give a general idea of the works, denote the size of the buildings, and show the disposition and arrangement of machinery.

In almost all the plans given, the machinery is arranged on the ground-floor only. This is convenient for the reception of the several engines and machines, and has great advantages where the different processes of manufacture are continuous from one department to another.

In the Bombay cotton mills, as already explained in note on cotton mill plans, the necessary light for working the mill is derived from a series of glass windows, supported on the *top* of iron girders of roof, the windows running right across the entire breadth of the mill from the rows of iron colums. The columns are placed generally about 20 feet apart one way, and 10 feet the other way, and these support the framing and top windows of the roof. For mills lighted on this

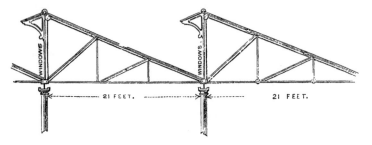

principle from the top several thousand square feet of window glass are required. Even for a cotton mill containing no more than 5,500 throstle spindles, which occupy much less space than mule spindles, more than 25,000 square feet of glass will be required. In the plans given, however, a different principle has been adopted for lighting, based on actual experience. It is a great improvement, inasmuch as it dispenses with the series of upper windows, does away with the iron columns, and with the nuisance of having several gutters inside the mill. The principal feature is this : that instead of one block, the factory buildings are divided into three or four separate wings, connected with each other, and leaving an open space in the centre for storing coals, building reservoirs, and such other purposes. The breadth of the wings is so regulated that the iron roof for each may be of one span, resting on the two side walls, and having gutters

outside. In one or two exceptional cases the breadth
exceeds 60 feet; in all other plans it is much less. In
those exceptional cases, if the breadth of the wings be
considered too great for a single span, a roof of two
spans may be adopted by having a row of columns
running in the centre line. The wings in any case for
all the plans given will be entirely lighted by a series
of windows inserted in the walls, to some extent lessen-
ing the cost in brickwork. No top windows, or even

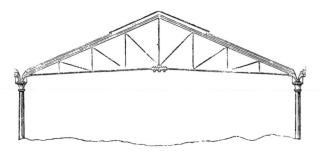

skylights, will be at all required. The windows being
placed opposite each other with top ventilators, and
the roof being of one span, the ventilation in the mill
could not be better; and the light that will be admitted
from side windows would be ample.

Working plans, with the necessary details, are given
by the machinist in duplicate after receiving the order;
these, being forwarded by mail steamer, and the buildings commenced forthwith exactly in accordance with
the plans, the factory will be nearly ready by the time
the machinery arrives from England. The work of
erecting the machinery and putting up the roof sometimes goes on simultaneously.

Extensions.—The plans have been so designed that extensions to works in a great many cases could be made without disturbing the machinery that may be working, at any time it may be found profitable to do so. It is a waste of capital and resources to erect factory buildings, as is often done, nearly twice as large as required in the first instance, and even to provide double the motive power, because future extensions may be contemplated. This is a very unsatisfactory way of proceeding.

The *construction* of factory buildings is, on the whole, as simple as building a dwelling-house, only it requires exactness in measurements. Generally speaking, the structure consists of four walls and certain partition-walls. The materials for building the factory will, to a great extent, depend on the market price. The walls may be either built of stone or of burnt bricks laid in mortar; the floor of stone or timber, with iron or timber roof. The thickness of a wall will depend on the height it is to be carried—the only condition required being that it should be substantial. Owing to the want of suitable machinery for sawing timber, and making wooden doors and windows, the cost is so high that iron windows made in Birmingham have been sent over for some of the Bombay mills; also iron framing for roof, and corrugated sheets of iron, galvanised, or of zinc.

Iron roofs for factories will be preferable in a great many cases to those of timber. The material is much more durable, dry, free from vermin, and not liable to be eaten up by white ants; nor is it so heavy-looking as timber. In one mill in Bombay, in a double-tile roof, to support the heavy weight, inch teak timber-planks have been used, costing more than £5,000, which item was entirely saved in another mill built close by, a few years later, by adopting the iron

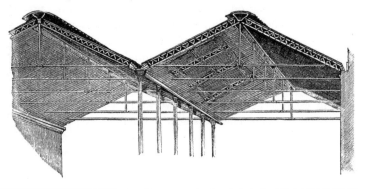

framing of roof and galvanised sheets without any boarding underneath. In one mill at Calcutta the iron roofs are double; that is, the framework has an outer and inner covering of galvanised corrugated iron, with a space of 12 inches between them, as a protection from heat. The principals of 40 feet span are formed of lattice-work rafters 10 inches in depth, dispensing with the ordinary struts and ties.

It is believed that, with a proper span of roof and good ventilation, there need not be any apprehension of inconvenience being felt even in the hot season in warm climates from iron roofs. In two mills in Bombay, one with a double *tile* roof with inch-teak boarding, and the other with only galvanised iron sheets *without* any boarding or felt, the temperature on a warm May day was the same in both cases; both mills were built, lighted, and ventilated on the same principle.

The more *fireproof* a factory building is the better it will be. This could be ensured to a great extent by avoiding timber as far as practicable—having brick arches resting on iron girders, doors and windows made of iron, and other improved appliances. It would be a wise provision to lay out, in all the principal rooms, water-pipes, and, at certain points, flexible leather or rubber hose pipes, with brass nozzles, in connection with the main town supply-pipe, to be used immediately in case of fire. In large works it would be wiser to have an independent steam fire-pump, so made as to be used for that purpose without any unnecessary delay. A steam fire-engine with double pumps, so made as to drive the machinery in a mechanic's shop, also with portable boiler and fittings, will cost about £150 to £200.

A *mechanics' shop* is required to be attached to every factory, for sundry repairs, fitted up with more or less tools, as already remarked in the section on iron work-shops. The best position for it in the factory buildings will be near the boiler-house; it should be driven by a small independent portable engine whenever required, which may also be so adapted as to work a fire-pump in case of fire.

Foundations for Steam-Engines, which are required to be very firm and stable, may be best built of stone, but not if stone is very expensive. In that case a solid brick foundation will answer nearly as well. By some mistake, for two cotton mills stone for engine founda-tions was ordered from England, and the freight came to more than 100 per cent. on the cost.

Firebricks and fireclay will be required to build round the boilers, to resist the great heat. In Madras, in the

School of Arts, firebricks are made, and have been used in the Government gun-carriage and other works there. Not so in Bombay, as for every mill firebricks and fireclay have been sent over from England.

The *Chimney* in factories is built of ordinary bricks. Iron chimneys will do well only for temporary, but not for permanent works, as a brick chimney is more economical in the end. It would be well to build the chimney 5 feet or more higher than actually required, as the draught is regulated at pleasure by the dampers, otherwise it will cost a great deal more, if increased in height after the works were completed.

The *Design* for a factory is not an unimportant matter. There has been, with hardly any exception, a total disregard of taste or design in mills built in India, at least in Bombay. Unfortunately a notion prevails that buildings with architectural beauty must cost a good deal; but not so necessarily. It is quite possible, without any additional cost, or at very slight extra cost, to relieve the hideous monotony and ugliness of a large brick surface extending hundreds of feet, by the introduction of wings and slight projections in suitable places, and by building arches over doors and windows of different forms. A pleasing effect is also obtained by using with some taste bricks of different colours. Of course in no case should the interior arrangement of machinery be unnecessarily interfered with, merely for outward show of the building. In England some mills are magnificent structures; for instance, that at Saltaire—as will be seen from the illustration—built by one who, once a farmer, by industry, enterprise, and intelligence, rose to be amongst the foremost in the rank of manufacturers in England. In

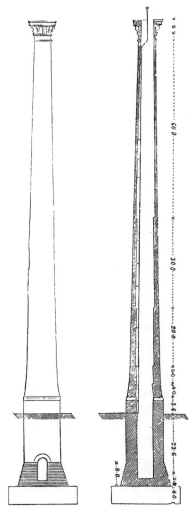

the French Exhibition of 1867 the English boiler-house was built from a design taken from an Indian mosque, of which a photograph and description had appeared in 1866 in the volume on the architecture of

Ahmedabad, published at the expense of a native gentleman of Bombay, who contributed £1,000 for this purpose.

English mills are generally five, six, and even seven stories in height, on account of the high price of land in towns, local taxes, and other causes; they are worked during the greater part of the winter season, even during the day, by gas-light, as the days are gloomy. In India this will not be necessary, on account of the superabundance of light; but when some particular mills are to be worked during night as well as day, a portable gas apparatus will be necessary, in which gas could be easily manufactured from coals or oils, or refuse fats. English mills, on account of the cold climate, are heated by steam-pipes during a greater part of the year; in India, and other parts of Asia, it will only be required in a very few places, on those days when the fall of rain during the rainy season may

be very excessive, and only for some particular manufactures.   In Bombay, where the rain-fall is 90 inches
during three months, on some days a few of the mills
are heated, but not all, on account of the excessive
moisture.   In Madras, and other places where the
amount of rain-fall is much less, no arrangement for
heating mills will be required.

   Cottages for the workpeople will be a suitable adjunct
to a factory—the more so where persons of various
ages and both sexes will have to be employed, as in
cotton, jute, and other large works.   The cottages
should be built within a very reasonable distance from
the factories, and fitted with hot and cold baths, and
other conveniences, and let at a nominal rental to the
workpeople.   The slight loss in the interest of capital
in building the cottages, the cost of which, compared
with the cost of factory buildings, would be insignificant, will be more than amply repaid by the advantages
gained.   It would tell even on the *financial* prospects
of the undertaking.

FACTORY BUILDINGS.

Particulars of Cost of a Cotton Mill at Bombay, covering 28,160 *square feet, of one story,* 14 *feet high, and lighted from the top. Given in rupees.*

	Rs.	Ans.	Rs.
Excavation for foundations, per 100 cubic feet . .	0	12	268
Excavation for 2 wells, 20 feet deep, per ditto . .	2	8	680
Rubble work for foundations, per ditto . . .	16	0	5,220
Upper rubble work, per ditto	20	0	1,056
Brickwork for walls, boiler, and engine-beds; chimney, 15 feet high; per ditto	31	8	16,622
Brickwork in chimney, height 80 feet, per ditto .	50	0	1,520
Brickwork in arches in engine-house, &c., per ditto .	36	0	798
Firebrick-work, labour only, at different prices, per ditto	0	0	330
Plastering the walls, in and out, per 100 superficial feet	6	8	3,584
(With well-made bricks plastering will not be required.)			
Stone pavement for flooring, per 100 square feet . .	35	0	8,354
Stone for shafting, size from 4·5 × 4·4 × 1·3, at different prices, per ditto	0	0	825
Dhopas and other stones for iron pillars, at different prices, per ditto	0	0	752
Teak doors, 13, windows, 20, for walls, per ditto .	0	0	1,642
(Upper windows, 640, of iron, not included.)			
Teak boards under iron sheets of roofs and brackets, 1 inch thick, per ditto	30	0	10,830
(This item has been saved in one mill by doing away entirely with boarding.)			
Miscellaneous timber-work, at various prices, per ditto	0	0	640
Iron roof and windows, including columns, girders, framing, galvanised iron sheets, gutters, &c., 640 iron windows, glass for windows, including freight, insurance, landing and carting, and charges for erection.	0	0	33,120
Firebricks and fireclay for boilers, including freight, &c.	0	0	1,424
Total rupees			87,665

PRICES OF IRON ROOFS, PER HUNDRED SQUARE FEET
OF GROUND COVERED.

Roofs of the ordinary ridge form, consisting of principals and other framing, skylights, gutter and rain-water pipes.

						£	s.	d.
In spans not exceeding 30 feet	.	.	.	7	5	0
,,	,,	40	,,	.	.	.	7	15	0
,,	,,	50	,,	.	.	.	8	2	6
,,	,,	65	,,	.	.	.	8	17	6
,,	,,	80	,,	.	.	.	9	17	6
,,	,,	100	,,	.	.	.	11	0	0
,,	,,	125	,,	.	.	.	12	12	6
,,	,,	150	,,	.	.	.	14	10	0

Corrugated iron roofs, in the arched form, without any framework (save tie-rods and suspension rods for counteracting the thrust upon the walls), are useful for spans not exceeding 40 feet.

					Painted.			Galvanised.		
					£	s.	d.	£	s.	d.
For spans not exceeding 25 feet, No. 20 gauge iron	2	15	0	3	10	0
,,	,,	33	,,	,,	18	,,	3	5	0	4	2	6
,,	,,	40	,,	,,	16	,,	4	2	6	5	0	0

Circular iron staircases for factories will cost (diameter 5 to 8 feet) from 18s. to 26s. per step.

Rough plate glass, cast, for factory windows, roofing, skylights, &c., will cost, ¾ thick in plates, from 20 inches long up to 120 inches, from 1s. to 2s. Chequered polished plate glass for office windows, &c., will cost from 1s. 6d. to 2s. 6d. per foot

Firebricks in England will cost per thousand, including square, arch, and chisel bricks, £4, delivered alongside a vessel. Fireclay, 26s. per ton.

Self-coiling revolving shutters for windows and doors, made of wood laths, connected with bands of steel or iron, or made in one sheet entirely of steel, the price ranges from 2s. to 4s. per superficial foot, and for factory gates and doors, &c., will answer better than heavy iron sliding-doors, being more secure and efficient.

Copper-rope lightning conductors are made in any length in one piece, and, according to the diameter, will not cost more than from 1s. to 2s. per foot.

NOTE ON FACTORY BUILDINGS.

The cost of factory buildings will be easily calculated from the dimensions given with each plan, in the place where the factory may be proposed to be erected. In Bombay the cost has been excessive, as will be seen from the figures given, though in several places it is believed it would be much less. However, this is a point on which it would be easy to arrive at a definite figure in a satisfactory manner by those most interested in the subject. If the walls are of brick, and the buildings of only one story, the thickness of the factory walls need not exceed 15 inches; the main wall, on which generally, in textile works, rests the main shafting, should be 18 inches; and the partition walls dividing one room from another may be 12 inches, or even less, in thickness. For a chimney, supposing it is 90 feet high, the thickness of the brick-work will be 18 inches at the base and gradually diminishing to 9 at the top.

The gauge of the galvanised sheet iron for roofs commonly used is No. 20; but this is governed to some extent by the spacing of the purlins, or longitudinal supports. Where these can be conveniently spaced at intervals of about 6 to 6½ feet, the No. 20 gauge is sufficient; but in the wider roofs it is often more convenient to employ stronger purlins at longer intervals, and in such cases No. 18 gauge is adopted. On the other hand, where the spacing is much reduced, No. 22 gauge will serve.

Zinc sheets, corrugated, and of Italian pattern (which is strong), have been used in India over iron framing for roofs. No boarding is necessary, but the Italian zinc is laid on rolls of wood, which fit under each flute of the corrugation. If felt is to be used to exclude the heat, it can be nailed down to the boards. Zinc perhaps has this advantage over corrugated galvanised iron sheets, that as old metal it will be always worth in India and the East, where it is used for making brass, nearly as much as it will cost in England. But it will cost more than corrugated galvanised sheets in the first instance. An iron roof, if not well put up, will be a source of great annoyance. Leak holes will be caused by rigidly nailing or screwing down each sheet of roofing in lieu of riveting the intermediate sheets, and fastening the ridge and eave ends. In one mill with half-inch teak planking, and felt underneath the sheets, on account of the sheets being screwed together, water came in in quantities; while in another, where the roof was well put up, and the sheets well riveted, without any planking or felt being used, not a drop of water leaked into the mill.

The size of the reservoirs for cooling the hot water from condensing or compound engines will have to be regulated according to the position of the ground; but the larger the surface exposed, and protected from the rays of the sun by planting large trees, the better it will be. The depth need not exceed 3 feet. In all plans given there is ample room for erecting large reservoirs in the yard, and thus the open space left will be economised for that purpose. If another reservoir is placed in front of the factory buildings with a fountain playing in the centre it will answer a double purpose. Where space is limited, large iron tanks for cooling the water are placed over the boiler house and other parts of the factory. In one mill there being two wells about 20 feet deep, one was made to answer for drawing the water for condensing steam, and the other for pouring in the hot water; the wells were thus used alternately. But the depth being the same, and the wells near each other, hot water from one communicated with the other, and also with other wells surrounding the mill property; as there was no cool water for condensing steam, there was great difficulty in working the engines; sometimes entire stoppage of the works was the result.

LEASE OF GROUND FOR BUILDING FACTORIES.

FORM OF INDENTURE.

This Indenture made .... between .... of ....
of the one part, and .... inhabitant, of the other part. Whereas
the said .... being absolutely possessed of, or well and sufficiently entitled to, an estate of inheritance in perpetuity of and in the
piece or parcel of land called .... and premises hereinafter
particularly described, hath contracted and agreed to, and with the
said ...., to grant to him a lease of the same for the term of
ninety-nine years from the first day of ...., 18 .., renewable for
ever, under and subject to the rents, provisions, conditions, and agreements hereinafter reserved and contained. Now this indenture
witnesseth that in pursuance of the said agreement, and for and in
consideration of the rents, covenants, provisions, and agreements hereinafter reserved and contained, and on the part of the said ....,
his executors, administrators, and assigns, to be paid, observed, and
performed by the said ...., doth, by these presents, demise
and lease unto the said ...., his executors, administrators,
and assigns, all that piece or parcel of land, situate at ...., and
entered into the books of the collector of land revenue under No. ....,
and containing .... square yards, be the same more or less, and
bounded on the north by the property of .... ; and which said
piece or parcel of land, or part, is more particularly delineated or
described on the plan thereof drawn on the margin of these presents ;
together with all trees of every description growing and standing
thereon, and all and singular the houses, out-houses, ways, paths, passages, water-courses, wells, drains, easements, privileges, profits, commodities, and appurtenances whatsoever to the said piece of land and
premises, or any part or parts thereof belonging or in anywise apper-
taining thereto. To have and to hold the same piece of land and
hereditaments, and all and singular other the premises hereby demised
and leased, or intended so to be, with their appurtenances, unto the
said ...., his executors, administrators, and assigns, from the
said first day of ...., for the term of ninety-nine years
thence next ensuing and fully to be ended and completed, yielding
and paying thereon unto the said ...., his heirs, executors,
administrators, or assigns, the yearly rent of ...., by equal ......
payments of ...., commencing from the first day of ....,
18.., free from all deductions whatever during the said period of ninety-

X

nine years. And the said .... doth hereby for himself, his
heirs, executors, administrators, and assigns, covenant and agree with
the said ...., his heirs, executors, administrators, or assigns,
that he, the said ...., his executors, administrators, or assigns,
shall and will, during the continuance of the said term hereby granted,
pay, or cause to be paid unto the said ...., his heirs, executors, administrators, or assigns, the yearly rent hereby reserved, at the
times, and in the manner, hereinbefore appointed, for the payment
thereof respectively, free and clear of all deductions whatsoever; and
he, the said ...., his heirs, executors, administrators, or
assigns, shall and will pay and discharge, or cause to be paid and
discharged, all taxes, rates, assessments, and impositions whatsoever
(except as hereinafter mentioned), now or hereafter during the continuance of the demise, to be taxed, rated, assessed, and charged, or
improved upon, any messuages, buildings, or trees, which may now or
hereafter be erected upon the said ground by the said ....,
his executors, administrators, or assigns. Provided always, and these
presents are upon this express condition, and it is hereby agreed
between the respective parties hereto, that if the rent hereby reserved,
or any of them, or any part thereof, shall be in arrears, or unpaid,
either in whole or in part, for the space of three months next after
any of the same shall become payable, then and so often as it shall
be lawful for the said ...., his heirs, executors, administrators, or assigns, to enter upon and distrain for the same, and for the
costs and charges occasioned by the non-payment thereof, upon all or
any part of the said hereditaments and premises, and to dispose of the
distress and distresses then and there found, according to law, to the
intent that thereby the said quarterly payments, and every part thereof,
so in arrears and unpaid, and all costs and charges occasioned by reason of
the non-payment thereof, shall be fully paid and satisfied, and further
that he, the said ...., his executors, administrators, and
assigns, shall well and truly pay, or cause to be paid, to the said
............, his heirs, executors, administrators, or assigns, the said
yearly rent of ...., on the days and at the times hereinbefore
reserved for payment thereof, without any deductions or abatements
whatsoever; and the said ...., for himself, his heirs, executors,
administrators, and assigns, do hereby covenant and agree with the
said ...., his executors, administrators, and assigns, that the
said ...., his executors, administrators, or assigns, paying the
said rent of ...., in manner aforesaid, and observing and performing all and singular the covenants, provisions, and agreements,
on his and their part to be paid, observed, and performed, as aforesaid, shall and may peaceably and quietly have, hold, and occupy,
possess, and enjoy the said piece or parcel of land, tenements, hereditaments, and premises, and all the trees, with their appurtenances, for
and during the said term hereby granted, without any let, suit, trouble,
molestation, interruption, or disturbance of it by the said ....,
his heirs, executors, administrators, or assigns, or any other person
or persons whomsoever, claiming or to claim by, from, or under him,
them, or any or either of them, or any of his ancestors, and that he,
the said ...., his heirs, executors, administrators, or assigns, or any

other person or persons whomsoever, claiming or to claim by, all taxes, assessments, and impositions upon or in respect of the said yearly rent, now or hereafter during the continuance of the demise hereby made to be taxed, rated, assessed, charged or imposed upon the ground hereby demised, or any part thereof, and that notwithstanding any grant, act, deed, matter, or thing whatsoever, by him the said ...., or any of his ancestors, made, done, omitted, committed, or executed, or knowingly or willingly suffered to the contrary, the said .... now hath in himself good right, full power, and lawful and absolute authority by these presents to demise, and grant, or secure of the said piece of land and premises hereby intended to be devised, and every part of the same, with the appurtenances, according to the true intent and meaning of these presents, and that free and clear, and freely and clearly acquitted and discharged, or otherwise well and sufficiently saved and kept harmless and indemnified by the said ...., his heirs, executors, administrators, of, from, and against all former and other gifts, grants, demises, assignments, leases, mortgages, charges, and encumbrances whatsoever had, made, done, committed, suffered, or executed by the said ...., or any of his ancestors, or any person or persons lawfully claiming by, through, under, or in trust for him, them, or any of them. And further, that he, the said ...., his heirs, executors, administrators, or assigns, shall and will, at the expiration of the term hereby granted, at the request, costs, and charges of the said ...., his executors, administrators, or assigns, and on payment of the sum of ...., by way of fine or premium for such renewal, forthwith duly make and execute unto him, the said ...., his executors, administrators, or assigns, a new and further lease of all and singular the premises hereinbefore demised, at and under the same yearly rent, and with and subject to the same conditions, covenants, provisions, and agreements, including this present covenant, as in these presents engrossed and contained. And further, that he, the said ...., his heirs, executors, administrators, and assigns, and every person whomsoever lawfully claiming any legal or equitable estate, right, title, term of years, or interest whatsoever into or out of the said premises, by, from, through, under, or in trust for him or them, or any of his ancestors (other than in respect of the said rents hereby reserved, and all the powers and remedies hereinbefore provided for securing and reserving the same), shall and will, from time to time, and at all times hereafter, on the reasonable request, and at the costs and charges in all things of the said ...., his executors, administrators, or assigns, make, do, execute, or cause and procure to be made, done, and executed, every such further and other granted deed or assurance in the law whatsoever for more effectually demising and assuring the said piece of land and premises unto the said ...., his executors, administrators, and assigns, according to the true intent and meaning of these presents, under and subject to the rents and covenants aforesaid, as by the said ...., his executors, administrators, or assigns, or their counsel in law, shall be reasonably advised and acquired. And, lastly, he, the said ...., for himself, his heirs, executors, administrators, and assigns, doth hereby covenant and agree with the

said ...., his executors, administrators, and assigns, that he, the said
........., his heirs, executors, administrators or assigns, will and shall,
from time to time, and at all times hereafter during the continuance of
any demise (unless prevented by fire or other inevitable accident), upon
every reasonable request, and at the proper costs and charges of the
said ...., his executors, administrators, or assigns, produce
and show in ...., to the said ...., his executors,
administrators, or assigns, or to his or their counsel, solicitor, or agent,
a bill of sale from the ...., dated ...., for the transpor-
tation, support, and defence of the possession, estate, right, title or interest of the said ...., his executors, administrators, and
assigns, of him to, or respecting the said land and premises hereby
demised and leased, or any part thereof; and also that he, the said
............, his heirs, executors, administrators, or assigns, shall
and will, upon the like request, and at the cost and charges of the said
............, his heirs, executors, administrators, or assigns, make
and deliver unto the said ...., his heirs, executors, administrators, or assigns, or to his or their counsel or solicitor, true and
attested copy or copies of and extract or extracts from the said bill of
sale.

In witness whereof, the said parties to these presents have hereunto
set their hands and seals, the day and year above written.

Signed, sealed, and delivered, by the within named ...., in
the presence of us.

GENERAL HINTS ON FACTORIES.

The estimates and prices given in this work, it need scarcely be said, must be considered as approximate. The prices of machinery will vary more or less according to the price of metals, wages, &c. Wages throughout all parts of England are rising steadily every year, and strikes for higher wages are common enough. Even in the same place, for precisely the same machines, the estimates (and exclusively from first-class makers) will vary from 10 to 15 per cent. or more. One maker may have orders on hand to last for some months ; another may be a little "slack;" or, if the machinery be for a new market, one who has more foresight, in order to obtain a hold in that market for his machinery, will not much care what percentage he makes. From this and other causes prices of machinery will vary even at the same place and same time.

Cheap machinery will prove the dearest in the end ; and it will be false economy to send orders to England to purchase such. It is not only necessary that machinery for India and the East should be made in a first-class style, of the best workmanship, materials, and finish, but that it should contain the latest improvements, or the improvements by which a greater quantity could be produced, or saving effected in the cost of production. This is of much importance, as the cost of the machine itself is a matter of secondary consideration, the progress of improvements in machinery being continuous. A new proprietor can avail himself to the fullest extent of new improved machinery, and beat the old ; whereas the old, even if he has the sense and desire, is not always in a position to avail himself to that extent. With the aid of improved machinery, obtained principally from England, and cheaper labour, without coals or even the raw materials, the Continentals, in some manufactures, compete with the English in several markets very successfully.

The cost of packing machinery, as a rule, amounts to 10 per cent. ; sometimes an additional charge is made for delivering it alongside the ship. Freight and insurance on machinery will amount to about 15 per cent., more or less, according to the class of machinery.

Erection of machinery is not at all a difficult task in a place where skilled labour is to be had, as each part of a machine, before being packed and shipped from England, is marked and numbered, so that it fits in exactly and truly when erected in the factory. When the machine is complicated, drawings and full directions are sent. The most important point in the erection of machinery is accurate level and exact measurements. For large textile factories, if trade is profitable, it would pay to get mechanics from England, and send them back to England after their work is completed. This is done even in Russia and other parts of Europe. At all events, for the present, for carrying on almost all the industrial manufacturing enterprises in India and the East, English managers or superintendents, or by whatever name they may be called, will be required, till a class of native mechanical engineers, trained up for the special work, springs up. The agreement with the European managers who are engaged in England is generally for a period of three years. It is clear that upon the right choice of a manager will depend in a great measure the success of the enterprise. The machinery may be the best, with all the latest improvements; but if the manager is an "incapable," or indifferent, or careless, the enterprise will prove a failure commercially. It is therefore absolutely necessary that the interests of a manager ought to be combined with that of the proprietors; and that, beyond receiving a fixed salary, the manager should receive a certain bonus out of the profits, or in proportion to the specific amount of work turned off beyond the stipulated quantity. Sometimes European managers, after arrival in the East, seeing that so much of the success of the enterprise depends on them, do things which they can never dare to do in England, and forget their duties. In such unfortunate cases, in order just to check them in time, and to remind them of their duties, all the necessary clauses should be inserted in their agreements; but the less the check applied, and the more they are treated with kindness, the better for themselves and their employers. A form of agreement has been given which requires merely the blanks to be filled up according to circumstances.

Native Mechanical Engineers.—Natives who have been sent to England, and trained as mechanical engineers, have done good service in introducing improved machinery, chiefly in Western India, and have received very handsome salaries. One was placed in a high position as superintendent of machinery in the Government dockyard, having under him several Europeans in the boiler, smithy, and other departments of machinery. But such examples have been very rare. In Bombay some of the cotton mills are entirely under the management of native engineers. But every native, before being placed in such a position, ought to receive a good English education, and a regular training in practice as well as theory, so that he may render efficient service.

Classes in the Indian Government colleges for civil engineering have been formed on the European model; and a native gentleman lately gave a large donation of money for a civil engineering college. But India wants mechanical engineers as well as civil, to assist in developing the resources of the country, hitherto so neglected; and it behoves Government to afford every facility to Indian youths for receiving the education necessary for a mechanical engineer. Lately a company was started in Bombay for advancing money at a very moderate interest to those natives who were desirous, and who were qualified, to proceed to England to receive their education as civil or mechanical engineers, and for other professions; but the company has collapsed, and nothing whatever has been done. The Government owes a duty to India in training her sons, who will be instrumental in acting as pioneers of civilisation, and adding to the wealth and prosperity of the country.

GENERAL FORM OF AGREEMENT WITH EUROPEAN MECHANICS.

ARTICLES OF AGREEMENT, made this ...... day of ...., One Thousand Eight Hundred and Sixty-...., between ...., residing at ...., of the one part, and ...., Engineer and Mechanic, of the other part,

Witnesseth, that for the considerations on the part respectively of the said ...., and the said ...., hereinafter expressed, it is hereby agreed between the said .... and the said ...., as follows, that is to say :—

1. That the said .... shall enter and be engaged in the service of the said ...., his executors or administrators, or any of his or their future partner or partners, successors or assigns, at certain works or mills of the said ...., about to be erected at ...., for the term of ...... years, to commence and be computed and terminable as hereinafter contained.

2. That in the discharge of his duties he shall, with all due care, attention, and diligence, conduct and complete in proper and perfect order, the erection and fitting-up and keeping in the most perfect working order and condition, all engines and valves, and all machinery, gearing, and shafting, as may be required of him, and all other machinery that may be requisite and necessary, to put the same into full and perfect work, and so to place and arrange all valves of engines that such engines may work steady and perfect, and so to set all valves connected with the several boilers that the steam will blow off at a given pressure, and so to fix such wheel gearing and shafting that the same will run evenly and work with perfect truth, and as is requisite and usual in all perfect machinery used in such factories, and that he will put the same into full work and take charge of and keep the same in perfect working order and condition, and so that they will not, nor will any of them, in any manner, spoil or injure any of the work or manufacture.

3. That he, the said ...., will remain in the service of the said ...., his or their executors, administrators, or assigns, or other the person or persons for the time being the partner or partners at any time hereafter in the said works of .... for the period of .... years from the date hereof.

4. That he will devote himself exclusively to the service of the said ...., or other the persons hereinbefore contemplated, in all the usual and regular hours of business, sickness or other unavoidable accidents excepted, and will not absent himself from such service without his or their leave or permission in writing first obtained, and will

obey all his and their commands and directions, and also all commands and directions of .... for the time being of such works, and to the best of his abilities will faithfully, honestly, and diligently execute such work as he or they shall desire him to do, and will also to the best of his abilities teach and instruct all the hands and apprentices employed in and about the said works in the art, mystery, and management and direction of all or any of the said .... engines, valves, boilers, and other the said machinery as may be required of him, and will also devote the whole of his time, attention, and abilities to the affairs and concerns of his said employers, and will do every act, matter, and thing required of him in reference to the concerns of the said business, and will also attend to every branch or department of such works as aforesaid, or which shall or may, from time to time, and at any time during the said period of service, be reasonably required of him, and will, in the performance of all such duties as aforesaid, use his best endeavours to promote the interest of his employers by due care and diligence, and will be faithful and just in all his dealings and transactions, and submit to all such rules and regulations in reference to business hours of attendance, leave of absence, and otherwise, as shall be in force and may from time to time be required to be observed, and shall in all things attend to and abide by the directions and orders of his said employer or employers or their chief manager for the time being, and will not himself, during the said term, nor permit others, to wilfully spoil, waste, embezzle, or destroy any tools, implements, machinery, or things committed to his charge or care or otherwise.

5. That if, during the continuance of the said term, through any stoppage of the works of the said ...., or if through any casualty whatever (not occasioned by or through the neglect or default of the said ....) occurring, by means whereof the said .... shall not be called upon to attend to the said concerns and business, the wages payable as hereinafter mentioned to the said .... shall continue to be paid notwithstanding, provided that if such stoppage or casualty be occasioned by such neglect or default then and during such time as the said works shall be stopped, the wages of the said .... shall cease.

6. That in the event of any such stoppage or casualty, from any cause whatsoever, and of the said .... being unoccupied, an account of the time so not employed shall be kept, and the said ...., or other the employer or employers, shall at any time thereafter be at liberty to require the services of the said .... for an equivalent period of time, but only at such reasonable hours beyond or over the usual hours of duty as he or they may appoint, and so that such lost or unoccupied time may be fully made up during extra hours; and it is expressly agreed that in respect of such extra time no allowance in the shape of extra wages to the said .... shall be made.

7. That he, the said ...., will not, during the said term of three years, engage, unless with the consent in writing of the said ...., or other the said person or persons for the time being constituting the principals in the said works, in any trade, business, or employment other than as aforesaid, nor will serve any other person or

persons for hire or reward in any capacity whatsoever or otherwise, nor attempt to impede his employers in their business, nor divulge any of their secrets or connections to any persons whomsoever, and will not do or engage in any act which may unfit or disqualify him from the regular and vigorous discharge of his duties, or which may be reasonably expected to induce or bring about any such infirmities or disqualification ; and that if he, the said ...., shall, without lawful excuse or cause, quit the service of the said ...., or of the person or persons interested in and managing the said concern during the continuance of these presents, such lawful excuse or cause to be certified in writing under his or their hands, the said .... will pay or cause to be paid to the said ...., his executors, or administrators, or other the person or persons aforesaid, the sum of .... sterling, and not by way of penalty, but as ascertained liquidated damages.

8. That in case the said .... shall be found to indulge in intemperate habits while in the performance of his duties, or in case of wilful negligence or inattention to directions, irregularities in attendance at the usual and customary hours of work, absence without leave, except as aforesaid, embezzlement, or not performing his duties efficiently or to the satisfaction of his said employer or employers for the time being, then the said ...., his executors or administrators, or other the said person or persons, shall be at liberty to deduct from the wages of the said .... such part thereof as may be equivalent to any loss he or they may sustain, and to discharge the said ...., on giving him one calendar month's notice to that effect, or on payment of one calendar month's wages, to discharge him immediately, and thereupon this agreement and all the clauses, matter and things herein contained, shall cease and determine, except so far as the recovering from the said .... the said damages as aforesaid.

9. That he, the said ...., will, during the continuance of these presents, find and provide himself with and pay all charges he may incur for board and otherwise.

10. And the said ...., on his part, agrees with the said ...., that in consideration of the said services of the said ...., and of the agreement on his part hereinbefore contained, he, the said ...., his executors or administrators, or other the person or persons for the time being interested in the said works, will, for the said period of ...... years, determinable nevertheless as herein contained, pay or cause to be paid to the said ...., the wages hereinafter declared, that is to say, during the first four calendar months thereof a salary of .... pounds sterling, or rupees .... per calendar month, and at the expiration of the said four months such salary shall be advanced and raised, and shall be at the rate of .... pounds sterling, or rupees ...., it being nevertheless declared that such last-named advance shall not be limited to the payment of the said .... pounds sterling, or rupees .... per calendar month, but immediately and so long as all the machinery in said works be running and kept in full and perfect work, then the said last-named advance shall be at once further

increased, and such salary shall be .... pounds sterling, or rupees
.... per calendar month, until the end of the said term of three
years, determinable nevertheless as herein declared; and it is also
declared that if at any time during the continuance of these presents
the said machinery, by being worked sixty hours during any week or
weeks, shall produce and turn off any extra quantity of the best or
first-rate quality over and above ...., then and in such event, and
while such extra quantity over and above the said .... shall be
turned off, the said .... shall be paid further wages over and
above and in excess of the said last-mentioned increase at the rate of
.... for every .... so produced or turned off over and above
the said .... ; it being distinctly understood and agreed that the
additional .... is only to be paid to the said .... in the
event of his producing the extra quantity of .... within the sixty
working hours named. And it is further declared, the said wages
of the said .... shall be paid monthly, the first sum of money
to be paid at the expiration of one calendar month from the date
hereof.

11. That the said .... will, at his own costs and charges,
convey the said .... as a second-class passenger by the overland route from England to ...., *via* Southampton; and should
the said .... faithfully observe and perform all the conditions
and stipulations herein contained until the expiration of the said term
of ...... years as aforesaid, or should these presents be put an end to in
writing by mutual consent, and no stipulation in such writing made to
the contrary, then the said ...., his executors or administrators, will at his or their own costs and charges, and without any expense
to the said ...., find and provide him with a second-class
passage back again to England. Provided, nevertheless, and it is
hereby mutually agreed between the said parties to these presents, that
in case the said .... shall remain in .... on his own
account after the expiration of the said term of ...... years as aforesaid,
or after these presents shall be put an end to as contemplated in this
clause, and does not leave .... within one calendar month
from either of the said periods, then the said .... shall not be
entitled to any money or assistance whatever on account of his travelling expenses home.

12. And it is hereby mutually agreed and declared between the said
parties to these presents, that if it be certified by two competent medical
men, one to be appointed by the said ...., and the other by
the said ...., that the climate of .... will prove fatal
to the health of ...., then the said ...., by giving
.... calendar months' notice in writing to the said ...., his
executors or administrators, of his intention so to do, shall be at liberty
to quit the service of the said ...., provided at the expiration
of the said term of .... months he forthwith proceeds to England at his
own expense and charges.

13. That in case the works, machinery, or any part thereof, be
wholly or partially destroyed by fire or otherwise, or if while landing,
or after the arrival and erection of the machinery in ....
aforesaid, the same shall become injured, it shall be lawful for, but not

compulsory upon, the said ...., his executors or administrators, at any time, to declare in writing to the said .... that this agreement is void and determined, and thereupon these presents shall cease and determine, except so far as contemplated in clauses 7 and 15 of these presents, and for the purpose of enabling the said ...., his executors, administrators, or assigns, from recovering the damages therein ascertained.

14. That if, in the event of the said ...., his executors or administrators, putting an end to these presents, as contemplated in clause 13 hereof, and there shall be nothing due or payable to him or them under clauses 6 and 13 of these presents, then the said ...., his executors or administrators, hereby agree forthwith to pay to the said .... three calendar months' salary from the date of such declaration, and to find and provide the said ...., without any cost to the said ...., a second-class passage home from .... to England. Provided, nevertheless, and this agreement is upon the express condition that such three months' salary shall not be paid nor payable, nor shall such passage be provided, unless the said .... shall forthwith when required return direct from .... to England as aforesaid.

15. And it is further declared and agreed that in case the said ...., his executors or administrators, shall find the said .... not competent to discharge his said duties in a skilful and efficient manner, and to the satisfaction of the said ...., or such other persons as aforesaid, or their chief manager for the time being, then and in any such case the said ...., his executors or administrators, may, if they shall think fit, put an end to this agreement, and every part thereof, on giving to the said .... months' notice of their intention so to do. And further, that in case the said .... shall neglect his said duties in any way, or shall absent himself as aforesaid without leave from the works or other place of business of the said ...., or from such other duties as aforesaid, as he or they, or their chief manager for the time being, may have directed him to perform, or committing any breach of the stipulations of this agreement, then and in any such case the said .... shall pay to the said .... the sum of .... as and for liquidated damages, and not by way of penalty, and it shall also be lawful for, but not compulsory on, the said ...., his executors or administrators, or others, the partner or partners of the said .... for the time being, or the chief manager for the time being, at any time within seven days from such neglect of duty or absence without leave, on committing any such breaches, to put an end to this agreement, and thereupon at any time during such seven days to discharge the said .... ; or, if the said ...., his executors or administrators, or other the said person or persons, shall not think fit to discharge the said .... within such seven days, it shall be lawful for him or them, but not compulsory, at any time within two months after such neglect of duty, to give to the said .... one month's notice of their intention to put an end to this agreement and every part thereof, and in such case this agreement shall from the expiration of that notice cease and be void, except so far

as the recovering from the said .... the said damages as aforesaid. And also that in case the said .... shall through any neglect or default of his own, or by reason of the breach of any stipulation or agreement herein set forth, occasion or cause any stoppage of any of the works of the said ...., executors or administrators, or his or their partner or partners, or any damage, loss, or expense to him or them, or embezzle, waste, or spoil any of his or their machinery, goods, or property, then and in such case the said .... shall, and he hereby expressly agrees to pay to the said ...., executors or administrators, or his or their partner or partners, as and for liquidated damages, and not by way of penalty, the sum of ...., and thereupon the said ...., executors or administrators, or the said partner or partners, may at any time thereafter, if he or they so think fit, but it shall not be compulsory upon him, or them, or their chief manager for the time being, at any time within seven days from such damage, loss, or expense, to put an end to this agreement, and forthwith discharge the said .... from such service as aforesaid, or that in case he or they shall not so put an end to this agreement and discharge the said ...., it shall be lawful for, but not compulsory upon him or them, at any time within two calendar months from the date of such damage, loss, or expense, to give to the said .... one month's notice in writing of their intention to put an end to this agreement, and every part thereof, and in such case this agreement shall, from the expiration of such notice, cease and be void, except so far as the recovering from the said .... the said damages as aforesaid. It being nevertheless expressly declared that should the said ...., his executors or administrators, his or their partner or partners, not put an end to this agreement as aforesaid, that he, the said ...., shall pay such damages as aforesaid, and that it shall be lawful for the said ...., his executors or administrators, his or their partner or partners, to deduct, from time to time, all and every damage sustained by him or them at any and every time from or out of any moneys that may then or thereafter be due and payable, or growing due and payable, to the said .... on any account whatsoever.

As witness the hands of the said parties the day and year first before mentioned.

(Signed)   ....

....

FACTORY WORKPEOPLE.

The Hands required for working in factories not being used to the factory system of working steadily a given number of hours, and being more accustomed to domestic labour, the change at first is not regarded with satisfaction by them; consequently the production for a short time will be less than otherwise would be the case. In England, at the commencement of the factory system, the regular discipline was not liked; so it will be in India and other countries where new industrial enterprises are to be conducted on the factory system.

Regularity of Attendance of the hands employed is one of the most important points for working with economy. A factory will require a certain number of hands to attend to the machinery; and the cost of fuel, superintendence, taxes, and other expenses will be the same, even if some are absent; but the production then will be less. In Bombay, where in nine mills out of ten great mismanagement prevails, there is much irregularity of attendance of the hands, and on some days even one-fourth of the machinery is not working from that cause; the wages in some cases being also in arrear.

But by giving reasonable wages, by introducing the system of wages according to the amount of work done, so far as it may be practicable, instead of paying a fixed sum by the month, by paying the wages regularly on a fixed day, by treating the hands with kindness, by building cottages or huts for them to reside in near the factory, and by taking other measures for their welfare, the irregularity of attendance, which is a drawback on the economy of production, will disappear to a great extent. In England the hands are paid regularly every Saturday. Even in England in most manufacturing towns there are a less number of hands present in the factories on Mondays than any other week days, as the men indulge too much in holiday-making on that day.

Prizes given to the workpeople every six months or every year, in the shape of cloth, saree, or other wearing apparel, for regularity of attendance, and for producing the largest quantity in a given number of hours, and for good behaviour, will be a welcome reward itself, and will stimulate their energies.

School.—Every factory where boys and girls are employed ought to have a schoolroom attached; and half an hour, mornings and evenings, should be devoted to giving lessons to the children in reading, writing, and simple sums. It will be found that after receiving this elementary instruction, order will prevail in the factory; they will be enabled to distinguish their numbers on the roll call, which will save time, and avoid confusion on the pay day; and they will attend to their work much better. It will improve their character and intelligence; by its influence their whole spirit will be moulded, if properly directed, and they will enjoy the blessings of reading and writing as long as they live. In England, by the Factory Act passed by Parliament, every boy and girl employed is compelled to attend school half the day. There are other provisions enforced by law for the health, comfort, and safety of the hands employed in the factories. A room also should be set apart for taking meals, if cottages are not built near the factory. Hot baths should also be provided by laying a pipe from the waste hot water of the engine.

Food.—Where cottages are built near the factory, as they ought to be, it will be better to establish a provision shop for selling rice, cloth, and other necessaries at wholesale prices to the workpeople. The proprietor will lose nothing by it, and for the slight trouble he will gain the affections and regard of the workpeople, which will be of some value at least. By adopting these hints, the cost to the mill proprietors will be very little or next to nothing, but it will be a good proof to the workpeople that their masters do care for them. They will work cheerfully; the production in the mill will be increased thereby, which will be a source of direct profit to the proprietors; and the men also will earn higher wages if they are paid according to the amount of work turned off. Thus the employers and the employed will be alike gainers.

In Great Britain, in other parts of Europe, and in America, many plans have been set in operation for the benefit of the factory work-people, which India and the East will do well to adapt to the habits and feelings of the workpeople, according to circumstances. One of the most complete establishments of this kind is in Yorkshire, at Saltaire, reared wholly by Mr. Titus Salt, the proprietor of most extensive spinning and weaving mills. Nearly 500 cottages have been built by him solely for his workpeople, and rented to them on very easy terms. A school has been also built, to give education to more than 600 children on the half-time system of the manufacturing districts of England. That is, half the number being engaged by turns in school, and the other half in the mill. The baths and wash-houses, with washing and wringing-machines, are complete. There is a medical man to look after the health of the people when they are sick, and there is not a single liquor shop in the whole place.

Another great mill proprietor, Mr. Akroyd, has built a village called Akroyden, with 112 cottages for the people employed in his spinning and weaving mills. He has also provided a number of allotment gardens, recreation grounds for swinging and for gymnastic exercises, and also a *penny* savings-bank. But such schemes are not merely confined to spinning and weaving mills. Messrs. Cowan, papermakers in Scotland, in connection with their works, have established schools, cottages, gardens, and even flower-shows, and a lodging-house for unmarried females; and they have taken pains also to encourage the factory people to make provision for old age.

At Price's Candle Company's Works, London, established for manufacturing candles, soap, &c., from palm oil,—one of the managing directors of that Company, Mr. Wilson, with the greatest zeal (and at first at his own expense), organised schools, cricket clubs, tea-parties, and even excursions during the holidays, for the improvement of the workpeople. At a meeting of the proprietors of that company, out of seventy present, it was agreed, with but one dissentient voice, to allow £900 a year for carrying on the educational operations, which had the desired effect on the workpeople to promote by their efforts the interests of the company. If some of the Bombay mills in the height of their prosperity, instead of voting, with great stupidity, large sums of money for marble statues, by the best artists of London, of directors, unenlightened, who did nothing whatever for the improvement of the factory or the workpeople, and who had made fortunes in shares of the company they had organised—if that amount, spent on marble busts, had been spent in building schools for the factory hands, the effect would have been very beneficial both to the workpeople and the proprietors.

AGRICULTURAL MACHINERY.

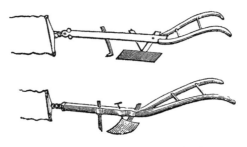

A practical farmer and a man of science stated, in one of the lectures on the Results of the Great Exhibition, that he fully endorsed the opinion given by the Exhibition juries, that the application of machinery to the main branches of farming labour, taken together, has effected a saving on outgoings or an increase on incomings *of not less than one half;* and that new agricultural machines have, with reference to the amount of saving produced by them, the merit of cheapness. These opinions come from a high authority, of unquestionable impartiality, and are the result of deliberate calculations based on actual practice, and are recorded in official documents. Let a claim be made then for these useful inventions for India and the East; as the landlord, the ryot, and the Government, have all an interest in the subject.

The condition of agriculture in India, where the Government derives *half* its revenue from land, is deplorable. Professor Wilson, member of the council of the Royal Agricultural Society, and Professor of Agriculture at the University of Edinburgh, and also a practical farmer, has remarked that the implement used for tilling the soil by the subjects of Queen Victoria in India at the present day is almost identical with that used for the same purpose by the Assyrians some three thousand years ago, as appears from drawings of ploughs taken from the tombs of Nineveh.

It has been stated that at the model Booldana farm, in Western India, the first step for improving the Indian plough has been taken, by substituting iron for wood at the joints, which cannot get loose, nor are so apt to break. But what is this improvement?—not even a drop in the ocean, compared with the gigantic strides made in America and England to facilitate the labours of the farm. Up to 1848 the number of patented inventions in the United States, solely belonging to the class of agriculture, was 2,043 ; and in the single year 1861, the number of applications was no less than 350, while in 1863 it was 502. There the profits resulting from the sale of agricultural implements are stated by a competent authority as enormous. One inventor sold the patent for a machine for threshing and cleaning grain for 60,000 dollars. Mr. James Howard, an English agricultural machinist, whose works are the largest in the United Kingdom for the special manufacture of field implements, after making a tour in America in 1866, in a lecture delivered before the London Farmers' Club, stated that the trade in America in agricultural implements had reached gigantic proportions, and that every year 100,000 reaping machines for cutting the harvest are made in the States. In England, where labour is cheaper than in America, according to the evidence of a practical English farmer and a man of science, by the use of the reaping machine, with two horses, corn is cut (fifteen acres in ten hours) at one third the cost, and in one twentieth the time required by the slow hand sickle ; it is prepared for the market by the use of the threshing machine, and the different qualities of the grain also separated by the machine at a cost of 7d. per quarter, against 3s. 5d. by the old process.

The progress in Great Britain in the manufacture and use of improved agricultural machinery is very rapid. Half a dozen firms made 327 threshing machines in the year 1852, but 1,084 in 1861, single houses turning out nearly 400 each per annum. The same number of makers sent out 270 steam-engines in the year 1852, but 898 in 1861. At the present time the estimated number of agricultural steam-engines at work in Great Britain is about 12,000; they are being supplied at the rate of about 10,000 horse-power per annum; and the average power of the farmer's engine is now 50 per cent. higher than it was ten years ago; this latter circumstance being due to the increased size and capabilities of the threshing machines, and also to the spread of steam tilling implements throughout the country. Four manufacturers sold only 32 reaping machines in 1852, but 1,715 in they ear 1861. Of these machines, so recently introduced, one factory has now turned out more than 5,000; and, including hay-mowers, four other houses have supplied altogether more than 20,000 in Great Britain, and half that number to other countries. One firm makes 4,000 root-cutters and pulpers every year; another firm makes 1,500 horse-rakes, 4,000 sets of harrows, and 7,000 ploughs per annum.

Even a machine for plucking cotton has been invented lately in America, which is expected to do the work of twenty men. The value of agricultural implements and machinery in use on farms in the United States in 1860, amounted, according to the official returns, to £50,000,000.

Now contrast this with India, with its area of 1,605,688 square miles, and a population of about 200 millions, where not one person is engaged in manufacturing a single improved agricultural implement, while in England one single firm turns out 7,000 ploughs per annum. There is not in India, throughout the whole length and breadth of the land, a single society, such as the Royal Agricultural Society of England, for the promotion of agriculture, nor an annual exhibition of improved implements, nor an annual show of cattle; nor is there a class formed in any school or college, for teaching agriculture on scientific principles, as in England.

In England and other parts of Europe, agricultural improvement has arisen, not only in making use of improved machines, but also in the manufacture and use of artificial manures. At one time, common farm-yard manure was considered as the only efficient fertiliser. But the astonishing effect of adding bone-dust to grass lands, of guano and various substances to other crops, even in very small quantities, has caused a complete revolution in agriculture. The demand for bones is so large that they are imported into England from abroad; and India also sends the bones of its dead animals from a distance of ten thousand miles to fertilise the soil of England, in ignorance of its use. In England, at first, pieces of bone of half an inch were used, then quarter inch, then fine bone-dust. The last improvement consisted in dissolving the bone in common cheap sulphuric acid (which is extensively manufactured in some parts of India); this was found to surpass greatly in efficiency the finest bone-dust; but English farmers found it inconvenient to prepare dissolved bones, so establishments were erected for that purpose, and the progress of this branch of industry has been so rapid that there are now in England and in other parts of Europe large factories solely devoted to manufacturing artificial manures, principally from bones, but also from certain mineral productions, refuse charcoal from sugar refineries, refuse salts from chemical works, woollen waste, &c. It has been found that many of these substances in their natural state, even after being pounded, have little or no effect on the crops, but that in order to restore the exhausted land, they require special chemical treatment.

The manufacture of artificial manures, in comparison with other branches of industry, is very simple, and does not require expensive machinery beyond the ordinary crushing and mixing apparatus, which are best set in motion by small steam-engines, or even by water power.

Centuries have rolled over India, yet the state of agriculture is the same. Let this be changed, as it has been in other civilised countries. Let the laws of nature be taught in Indian schools; then the Indians will seek for scientific principles, and not be content with the observance of mere routine. Let science lend her aid in Asia in relieving the necessities and in advancing the comforts of its inhabitants. By such change, upon no class will greater benefits be conferred than upon the agricultural; the landlord and the Government will also share in it, as half the revenue is derived from land; and the manufacturer in India and England will rejoice, as what benefits one must benefit the other. The deeper the plough will go in the soil in Asia, the faster the shuttle will fly in the weaving looms.

PRICE LISTS OF STEAM AGRICULTURAL MACHINERY,

ADAPTED FOR LETTING ON HIRE, OR FOR USE BY LARGE
LANDED PROPRIETORS.

£

Steam cultivator, with ploughing apparatus, windlass, steel
    ropes, with portable steam-engine, mounted on wheels,
    complete, with harrows combined, for working on the single
    engine system, from £600 to   800
    For working on the double engine system, from £800 to .   1,500

The above cultivator effectually cuts up land at one operation, and
also prepares the ground already turned up, effecting two operations at
one time if required. Without the removal of the engine or windlass, as much as 40 acres can be cultivated, or 7 to 10 acres in a day, on
the single engine system, and the engine can be used for threshing or
other purposes.

£

Steam threshing machine, fitted to prepare the grain for market,
    mounted on wheels complete, driven by straps, from £150 to   500

The operations of this machine are threshing, winnowing, straw
shaking, and delivering cleaned corn into bags at the end of the machine. They are suitable for places where grain has hitherto been trodden
out by the feet of oxen, and will thresh from 24 to 68 bushels of corn.
Some machines are fitted with a straw chopper, which reduces it into
small pieces.

£

Steam centrifugal pumps, to raise water for irrigation, 10 to 20
    feet high, from 500 to 4,000 gallons of water per hour,
    from £75 to   280
Disintegrator for pulverising boiled bone, bone ash, guano, and
    other materials for manures, with casing complete .   .   80

(Extracts from the Report on Steam Cultivation.)

" We have seen one instance where good results were obtained on 138
acres, but the circumstances are too exceptional. We think 250 acres
of strong arable land is the minimum quantity on which it would be
wise to introduce steam culture, the engine still earning most of its
money at other work; in such a case we decidedly recommend the
roundabout plan, with a cultivator and plough. . . . On lighter
land a larger breadth would be desirable, say from 350 to 400 acres,
and as the acreage is increased beyond these limits, the profit of the
investment would be increased. . . . Farms require more or less
preparation for steam; fences should be taken up, and in some cases
roads made. . . . Trees left in the arable fields present a serious

obstruction. . . . In the majority of instances we found the proprietors satisfied with results, and, having once experienced the advantages of steam over horse-power, unwilling to go back to the old
system. . . . A point of importance on strong land is the effect of
steam cultivation on drainage and produce. . . . In many cases
the increase has not been sufficiently marked to be visible to the eye,
whilst in others from 4 to 8 bushels per acre is the estimated increase
of corn crops, and such a result would add materially to the profits on
steam."—From the *Journal of the Agricultural Society of England*, July,
1867.

AGRICULTURAL MACHINERY FOR ANIMAL POWER.

	£
ron plough, with all recent improvements, made suitable for India and the East, from £6 to	15

These ploughs are manufactured with slight modification in details
for heavy or light soil, for working at various depths; or the various
processes of ordinary ploughing, paring, ridging, and subsoiling, can
be performed by one implement. The average weight of these ploughs
is from 2 to 3½ cwt.

	£
Potato raising plough, adapted for places where large crops of potatoes are raised. From 3 to 4 acres can be raised in a day with two horses. From £6 to	8
Dwarf iron plough, for shallow ploughing, worked by a bullock. From £3 to	5
Iron harrow, two to four beams, five or six rows of teeth, for general seed harrows; weight from 1 to 2¼ cwt. From £6 to	8
Double-action hay-making machine, weight from 7½ to 12 cwts. From £14 to	20
Threshing machine, complete, for being worked by bullocks, from £56 to	100
Grass cutting and reaping machines, with all improvements, from £20 to	30
Pumps for irrigation, to be worked by bullocks, and to raise 4,000 gallons per hour	150

APPENDIX.

THE TEXTILE MANUFACTURES OF INDIA.

(Extracts from Dr. Forbes Watson's work, as originally printed for the India Office, 1867.)

Specimens of all the important textile manufactures of India have been collected in eighteen large volumes, forming one set containing 700 specimens. . . . . This work, therefore, may be regarded as an analysis of the contents of the eighteen volumes, and a classification of them according to function, quality, material, and decoration.

Muslins.—A large proportion of these, and certainly the most famous of them, are manufactured at Dacca. Other places in India produce fabrics of extreme delicacy and beauty, though the Dacca weaver has unquestionably the first place, having never as yet been beaten, either in India or out of it. No one will examine them and marvel that they should have received such poetic names as the "Evening Dew," and "The Running Water," and the "Woven Air." The weight of a piece one yard wide and four yards long, was found to be 566 grains, and the weight of another piece of the same width, but ten yards and twelve inches long, was found to be 1,565 grains.

Jamdanee, or loom-figured muslins, from the exquisite delicacy of manipulation which many of them display, may be considered the *chef-d'œuvre* of the Indian weaver. From their complicated designs, they have always constituted the most expensive productions of the Dacca loom.

Calicoes.—The common unbleached fabrics, under names varying in different localities, constitute a large proportion of the clothing of the poor. They are also used for packing goods, and as a covering for the dead, for which last purpose a large quantity is employed by Hindoos and Mohammedans.

Canvas, Cotton.—The strength, lightness, and other good qualities of the cotton sailcloth manufactured in India, recommend it to more attention than it has received in this country. The quantity of cotton annually consumed in India in the manufacture of sail and tent cloth is very large.

Checks and tartan patterns, made with cotton woven with coloured thread, are admirable imitations of well-known patterns in this country. They are chiefly used for skirts, petticoats, &c. Some of the shepherd tartans are also used for making up into trousers.

Dyed and printed fabrics are produced in many parts of India, or perhaps more properly speaking, here and there over the country. Musulipatam, Arnee, and Sydaput, in the Madras Presidency, are famous for their *kheetee*, or *chintzes*. Those of Musulipatam are known under the name of *calum-kouree* (which literally means "firm colour"), and exhibit great variety in style and quality.

Cotton, Miscellaneous.—Not a few of them, such as the table napkins, d'oyleys, and pocket-handkerchiefs, are manufactured to suit European wants, and these illustrate the imitative power of the native manufacturer.

Silk and Cotton piece goods form an extensive article of manufacture in many parts of India, chiefly for home consumption, but partly also for export. An important class of fibres, commonly known under the name of Mushroo, is a satin with a cotton back. It is a favourite material, and is used in a variety of ways by the well-to-do classes. All Mushroos wash well, especially the finer kinds. English or French satins are more beautiful both in colour and texture ; but it is needless to say they will not wash.

Wild Silks—in contradistinction to the foregoing, or cultivated variety—are the Tussar, Eria, and Moonga, and fabrics made of some of them—and particularly the Moonga—have probably been known in the East from time immemorial. Although Tussar is the variety of wild silks best made in this country, the Moonga, from its superiority in point of gloss and other qualities, is that most commonly employed, especially for the manufacture of mixed fabrics, and for some kinds of embroidery. The silk Moonga is imported into Dacca from Sylhet and Assam. The cloths of this class are of considerable variety, both as regards texture and pattern. Some consist chiefly of cotton, with only a silk border, or a silk flower or figure in each corner ; others are striped, chequered, or figured with silk throughout the body of the cloth. These cloths are made exclusively for the markets of Arabia. Some are occasionally shipped to Rangoon, Penang, and places to the eastward, but the far greater portion of them is exported to Jidha, whence they are sent to the interior of the country. A considerable quantity of them is sold at the annual fair held at Meena, in the viciniy of Mecca. They are made into turbans, gowns, vests, &c., by the Arabs.

Kincob.—Of the varieties and patterns produced in India by the combinations in the loom of silk, gold, and silver, only a faint idea can be obtained from the specimens. The European manufacturer who may have attempted the introduction of metal into his fabrics, will all the more readily comprehend and admire the results obtained by the Indian weaver.

Gold and Silver Tissues.—In these tissues the flattened wire, instead of being twisted round silk thread, is itself used : the warp, or the weft, as the case may be, being of very fine silk thread, so as to interfere as little as possible, with the continuity of the surface presented by the metal. It is thus that the *cloths of gold and silver*, of which we hear in Eastern countries, are made.

Hand, or Needle Embroidery, is a kind of work in which the natives show an admirable skill. Every kind of fabric, from the coarsest muslin to the richest cashmere cloth, is thus decorated; and though Dacca and Delhi are the places best known for their embroidery, there are numerous other places in India in which the workers are equally skilful. *Chikan work* includes a great variety of figured or flowered work on muslin, for gowns, scarfs, &c. It also comprises a variety of network, which is formed by breaking down the texture of the cloth with the needle, and converting it into open meshes.

Cashmere Shawls.—This is now by far the most important manufacture in Punjaub, but thirty years ago it was almost entirely confined to *Cashmere.* The best shawls in the Punjaub are manufactured in Umritsur, but none of the shawls made in the Punjaub can compete with the best shawls made in Cashmere itself. A woven shawl made at Cashmere, of the best materials, and weighing seven pounds, will cost in Cashmere as much as £300. Of this amount, the cost of the material, including thread, is £30, the wages of labour £100, miscellaneous expenses £50, duty £70. . . . . *Pushum,* or shawl wool, is a downy substance, found next the skin, and below the thick hair, of the Thibetan goat.

Camel's Hair Cloth, called *puttoo,* is usually considered to be manufactured from the inferior qualities of shawl-wool. The puttoo is generally employed by the natives for making up into long coats, called *chogas,* ornamented in a variety of ways, generally by means of silk braiding.

Kerseymere-like Cloths, unlike the puttoo, are of a rather hard description, like our kerseymere cloths.

Woollens, striped, are made up for wear in Sikkim, as well as in Nepal and Thibet. *Cumblee* is worn in the cold weather for the protection of the head and shoulders. Felts are used for blankets and cloaks, and for making into leggings. These felts are commonly used as carpets, cushions, bedding, horse clothing, &c.

Carpets and Rugs are of five kinds, and the manufacture is of very considerable extent. The first is made entirely of cotton, and is of a close, stiff texture, and smooth surface. The ordinary name of these is *Suttringee,* and they may be said to be made here and there over the whole country, their use being almost universal. They are extremely durable. In the second kind the warp, like the last, is of cotton, but the woof is of wool. The loom employed in weaving both these is horizontal, without either treadles or reed, and the warp is stretched out the whole length and breadth of the piece intended to be wrought. The woof is not thrown across with a shuttle, but is passed through by several workmen, who bring the threads together with wooden combs in place of a reed. The narrowest piece requires two men, and eight or ten men are employed when the breadth is great. The third kind is made of cotton, but instead of presenting the plain surface of the two last, a short, thick-set pile of cotton is worked into it. In the fourth group the pile is of wool.

PLACES OF MANUFACTURES IN INDIA.

NAME OF PLACE	PRESIDENCY OR DISTRICT	NAME OF PLACE	PRESIDENCY OR DISTRICT
Agra	N.W. Provinces	Hoshiarpore	Punjab
Ahmednugger	Bombay	Hyderabad	Sind
Arcot	Madras	Hyderabad	Deccan
Bangalore	Mysore	Jegpore	Native State
Beckaneer	Rajpootana	Jhelum	Punjab
Beejapoor	Suttara	Kangra	Ditto
Belgaum	Bombay	Karicul	Tanjore
Bellary	Madras	Kohat	Punjab
Benarus	N.W. Provinces	Kurnool	Madras
Berhampore	Ganjam	Lahore	Punjab
Berhampore	Moorshabad	Leiah	Ditto
Bhagulpore	Bengal	Loodiana	Ditto
Bhairulpore	Native State	Madras	Madras
Bhurtpore	Ditto	Madura	Ditto
Bickul	North Canara	Mangalore	South Canara
Bombay	Bombay	Musulipatam	Madras
Broach	Ditto	Moorshedabad	Bengal
Buttala	Goodnapore	Mylapore	Chingleput
Cachar	Bengal	Nagang	Madras
Calcutta	Ditto	Nagpore	Nagpore
Cashmere	Native State	Nellore	Madras
Chicacole	Gangaum	Nurrapore	Sind
Chingleput	Madras	Odeypore	Rajpootana
Chunderee	Gwalior	Palamcottah	Tinnevelly
Coimbatore	Madras	Patna	Bengal
Combaconum	Tanjore	Pondicherry	Sonth Arcot
Congeveran	Chingleput	Pulicat	Madras
Coonathoor	Ditto	Radnagore	Ditto
Cuddalore	Arcot	Rajamundry	Ditto
Cuddapah	Madras	Raneepore	Ditto
Candapore	South Canara	Rawul Pindee	Punjab
Cutch	Native State	Salem	Madras
Dacca	Bengal	Santipore	Bengal
Darjeeling	Ditto	Sattara	Bombay
Delhi	N.W. Provinces	Shahabad	Patna
Deyra Ishmail	Punjab	Shikapore	Sind
Dharwar	Bombay	Sikkim	Native State
Futtygurh	Furracabad	Surat	Bombay
Gangam	Madras	Sydaput	Chingleput
Goodaspore	Punjab	Tanjore	Madras
Gwalior	Native State	Trichinopoly	Ditto
Gya	Bengal	Vizagapatam	Ditto
Harzara	Punjab	Warungul	Deccan

MINERAL RESOURCES OF INDIA.

(Extracts from a Paper " On the Results of the Great Exhibition, 1862,"
in the *Practical Mechanics' Journal*, by M. C. COOKE, Esq., F.S.S.,
India Department, London.)

We are only now beginning to understand in what the mineral
wealth of India really consists, and it is gratifying to learn that the
coal and iron resources have so increased in development as to occupy
very prominent positions in this collection.

Iron.—The most important part of metalliferous deposits is the iron
series. Of these ores a good collection is shown from various localities.
The red ochrey ironstone of Cuttack is represented by specimens from
Dhenkenal and Talchere. An abundance of this ironstone is found in
the district of Sumbulpore, and it is plentiful in the Cattack tributary
states of Talchere, Dhenkenal, Pal-Sahara, and Ungool, and, in fact,
throughout the hilly country bordering the settled districts on the
north-west. All the iron employed in this division is obtained from
these local sources. In Sumbulpore the crude iron is sold for about
three farthings per pound. In smelting the ore no flux is used. The
broken ironstone is mixed with charcoal, and put into a clay furnace of
about four feet in height. The fire is maintained by an artificial blast,
introduced through a fire-clay pipe, which is luted with clay after the
insertion of the nozzle of the bellows. The slag is raked out through
an aperture made in the ground, which runs up into the centre of the
furnace base. The charcoal employed is that of the Saul tree, which is
abundant. Although limestone in calcareous nodules is plentiful on
the spot, it is nowhere used in smelting. A specimen of the Ungool
ore taken from the ground, where it had lain exposed to sun and rain,
gave 66 per cent of teroxide of iron, equal to 46 per cent. of metallic
iron. A sample from Pal-Sahara gave $60\frac{1}{2}$ per cent. of protoxide,
equivalent to 47 per cent. of metal. Iron ore, of which a specimen is
exhibited, is found in the vicinity of Moonghyr, in the Korruckpore
hills, and is smelted by the natives for local use.

The whole chain and spurs of the Vhyndhya range, in the Shahabad
district, is full of mineral stores. Abundant quarries of peroxide and
protoxide of iron are opened in the accessible portions of the Kymore
range, which is a spur of the Vhyndhya. Most of the ores are rich in
metal, some of them yielding from 70 to 75 per cent. of pig iron.
Some of the best iron in India is produced at Palamow and Singrowli.
The Singrowli iron, especially, bears a high character in the market, it
being tough, flexible, and easily worked. The greater portion of the
Kymore ores are found on what is termed the old red sandstone, super-
lying fossiliferous limestone of indefinite thickness. Although there is
an abundance of mineral coal in South Mirzapore, in Palamow, and

Singrowli, the native smelters only employ wood charcoal, and the whole process is conducted in the simplest manner.

The East Indian Iron Company exhibits a very interesting series of iron ores, accompanied by articles of manufacture. The ores are those worked by the company, and consist of samples from Salem, South Arcot, and Beypore; the stone of which the shell of the blast furnaces are built; the shells employed as a flux; samples of wood used for making the charcoal employed in smelting; native goat skin bellows; charcoal pig iron, bar iron, cast iron, and Bessemer steel, made at Beypore, direct from the furnace, with specimens of cutlery, made by native smiths from the above-named Bessemer steel, and cutlery made from the same steel at Sheffield. This series is included for convenience and comparison in one case, and is as important as it is complete and interesting.

The iron ore of the Salem district is a rich magnetic oxide, which is very heavy and massive. The yield averages 60 per cent. of metal. A portion of this ore is a pure black magnetic oxide, which would yield as much as 73 per cent. of iron. The ore is sometimes much mixed with quartz, which is a very refractory material in the blast furnace. The chrome ores of Salem are rich, but have hitherto been turned to little account.

The Kumaon Iron Company furnish samples of their metal, but no information as to the kind or source of the ores employed.

Specimens are also sent by his Highness the Maharajah of Gwalior of the *dhaoo* or iron earth found in the Gwalior district. Iron is obtained from this ore at the cost of twelve annas for twenty seers weight, or eighteenpence per 40 lbs. Ores are also shown from Tendookhera and Agureea. At Tendookhera the ore actually worked is a large vein in the limestone of the great schist formation of the Indian Survey. It occurs to the north of the Nerbudda, in the open flat country between the river and the Vhyndhya hills. Only one mine is worked at present, but a similar ore has been found at other places in the neighbourhood. The ore resembles that of the forest of Dean, and contains about 40 per cent. of iron. It is calcareous and very fusible, and is largely smelted by the natives at Tendookhera, where about sixty furnaces are generally at work. The ore is obtained by means of pits sunk from 30 to 40 feet through the alluvium of the valley. From the iron of these mines the suspension bridge of Saugor was built several years ago. The Agurea ore is obtained in thin flakes of a grey colour and metallic lustre. The mines are situated on a hill consisting of iron ore found at eighteen inches from the surface, extending over an area of 60,000 square yards, with an average depth of thirty feet.

Petroleum.—Petroleum is sent from Burmah, but although the supply is unlimited, the price is high, as the monopoly is in the hands of the king. A similar product is forwarded from Assam, but without more definite information of the precise locality. A kind called Cheduba petroleum is sent from Akyab, where it is used by the natives for burning in lamps, and varnishing the bottoms of boats.

Plumbago.—That hitherto obtained from Indian localities has scarcely been of a quality sufficiently fine for pencils, but would be available for crucibles : the demand for it for such purposes being on the increase, it will not be uninteresting to learn that specimens have been sent from a mine discovered last year at Goorgaon.   It is found in masses of variable sizes, and in general quite detached.   In some cases the rock all round is full of plumbago, mixed with small micaceous particles. Provisions are being made for a further and more minute examination of this deposit.   This mineral is also contributed from Vizagapatam and Malacca.

Sulphur.—Sulphur is also forwarded from two or three localities, amongst which may be named the Hala mountains, west of Gundava, and the mines near Shorun in Beloochistan.

Amber.—There are some manufactured articles of amber of excellent quality, but the crude material sent is unworthy of notice.

Rocks or Building Stones.—Porphyritic and felspathic granites are shown from Shahabad, and limestones from Jubbulpore, Chota Nagpore, Cheynepore, and the Rohtas range.   A small specimen of an excellent alabaster is exhibited from Jacobabad, and a quantity of rock formations without locality or information of any kind, the papers accompanying them having apparently been mis-sent.

Amongst the slate series are some hone-stones from Chittledroog and North Arcot, and what is known locally as *Moongee* stone, which seems to be a kind of chlorite slate or Potstone.   This stone, when quarried, is soft and easily worked, but soon becomes hard, black, and bright. It is locally employed for the manufacture of utensils, idols, and carved figures generally.   It is extensively quarried at Arissa.

Corundum and Emery.—Those exceedingly useful minerals, corundum and emery, especially the former variety, occur in at least thirty different localities in the Presidency of Madras.   Excellent qualities may be procured from Salem, Guntoor, Nuggur, and Coimbatore. Both kinds are frequently found associated together, as at Salem, Nuggur, and Nellore.

Clays.—Pottery clays are abundant in Bangalore, and although samples of the clay have not been sent, there are a few typical specimens of the pottery produced from them.   A pure porcelain clay, convertible into a good white ware, is found in great quantities in Bangalore, and may be had for the trouble of picking it up.   A buff clay of good quality, and a brown potters' clay, abound.   These, combined with felspar grit, are now manufactured into drain pipes, &c., by the prisoners at Bangalore Industrial Convict School, which establishment has been in existence only about six months; and one of its chief objects seems to have been the utilisation of the vast quantities of clays which, though so abundant, had never before been turned to account.

Excellent clays abound, of which the following are amongst the best :—White plastic clay, resembling the ball clay of the English potter, may be procured at Chingleput, Palaveram, Conjeveram, Cuttapaukum, Cuddalore, Cuddapah, Coringa, and the Neilgherries.   Tough yellow clay is plentiful on the Red Hills, Poonamalee, Chingleput, and Cochin.   Greyish-white clay at Bangalore, Tilaveram, Sireepermatoor, and the Neilgherries.   Red clays at Burmah and the Guntoor districts. True fire clays at Chingleput, Cuddapah, and Tripasore.   Clunch at

Madras, Poonamallee, &c.   Examples of these from some of the localities are exhibited by Dr. Alexander Hunter of Madras, and others.
Specimens of black and white clay are shown from South Arcot and
the Nizam's territories.   These are employed locally for the manufacture of pottery.   Fire clays, brick clays, and clays for hydraulic
cements are forwarded from Singapore.

Kaolin may be obtained readily at Arcot, Hyderabad, Bimlapatam,
Cuddapah, Bangalore, Chittoor, Madura, Cochin, &c.; but this substance
is very little used in India, on account of the ignorance of the natives
as to the best methods of combining and firing it.

Pigments.—Of miscellaneous contributions not coming within the
foregoing groups there are palæontological and other specimens from
the survey under the direction of Professor Oldham; earths employed
as pigments, such as chalk coloured by the presence of oxide of iron;
fullers' earth, gypsum, sulphate of barytes, chalk, grit, chert, and
sandstone, including flexible sandstone from Darjeeling; mica, talc,
and materials employed for the production of lime; and lastly, the
saline earths and chemical products obtained from saline efflorescence.

Saltpetre and Sulphate of Iron, the former obtained from a saline
efflorescence, and the latter from the mineral sulphate of the Kymore
range.

Shale.—Specimens are exhibited of a black shale found principally at
Kalabag on the Indus, from which alum is manufactured by a process
very similar to the English method.   An average of 430 tons are
annually sold of the alum produced from this shale.

ARTS, MANUFACTURES, AND MACHINERY AT THE PARIS EXHIBITION OF 1867.

(Extracted from the *Engineering* and other English journals.)

The present is essentially the age of machinery, a comprehensive term which includes every form of productive apparatus. A machine is to a tool what an organ is to a whistle, or a regiment to a private soldier—a harmony of forces, a well-arranged combination capable of producing great results. The peculiarity of the time is vastness. In the Paris Exhibition you may see the mightiest preparations made for accomplishing the most ordinary and seemingly common-place results. The reflections with which one comes away from a walk among the machines in the outer ellipse are that the curse of labour is at length about to be removed from long-suffering humanity, and a new era begun in our history when the sweating brow shall give place to the attentive brain, and brute matter only require man's superintendence while it subdues itself.

Cotton Gins.—As makers of spinning and weaving machinery, Messrs. Platt and Co., of Oldham, have a reputation second to none. In separating the cotton fibre from the seed, they exhibit three gins on Macarthy's principle, in which a roller of a few inches diameter, built up of fibres of jute in the manner of a brush, but so compressed as more nearly to resemble timber or pasteboard, is slowly rotated on the front of a hopper, into which the uncleaned cotton is thrown, where it rests upon a wire grating sloped towards the roller, but not quite reaching thereto. In front of the mass of uncleaned cotton a long horizontal blade reciprocates against the horizontal jute roller, an opening being made in front of the roller to permit this action to take place. The fibres of cotton are caught between the vibratory blade and the slowly rotating roller, and as the blade descends the fibres are pulled away from the central seed, which, when denuded from the enveloping fibre, is small enough to fall through the narrow opening formed by the grating not being brought quite up to the roller and blade. The seeds thus collect in the lower part of the hopper, from whence they are easily removable. This gin is more suitable for cotton of short or tender staple than the saw gin of Whitney. In *cotton spinning* machinery, improvements have been so great, and the difficulties in the way of a satisfactory performance of the work have been so well overcome, that, simply by being more carefully opened, drawn, and doubled, counts of yarn are now being spun from cotton, which a few years ago was not thought capable of being applied to the trade.

Sewing Cotton.—Among the houses who have distinguished themselves by adopting all improvements, and creating a large trade, we must name that of J. and E. Waters and Co. Their trophy is wholly composed of sewing cotton on spools, cards, and in balls; these latter are extensively manufactured for India, and other foreign markets.

Oil, Soap, Candles.—At the present time stearine is manufactured by three processes: 1st, by saporification of the fatty matters with 13 or 14 per cent. of lime, thus forming soaps—stearates, oleates, and margarates of lime—insoluble in water, whilst the glycerine set free is dissolved in the water; 2nd, by treatment with sulphuric acid, followed by distillation. The third process consists in the decomposition of the fatty matters by the combined influence of water, heat, and pressure. It possesses the advantages of the first without the disadvantages of the second. It is economical, and produces a white stearine, and, at the same time, an acid oil of good quality; but there is the difficulty attending it that the treatment of the fatty matters subjected to it should be carried on in a vessel sufficiently strong to resist with safety an internal pressure of about 210 lbs. per square inch. A vessel of this kind for containing the fatty matters whilst being treated under pressure cannot be constructed of iron, on account of that metal being rapidly attacked by the fatty acids, and it has therefore to be made of copper, a material, however, which is not so well adapted for resisting the pressure as iron, particularly at the temperature at which the process has to be carried on. A plain copper vessel for the purpose we have mentioned has therefore to be made of considerable thickness, and it is therefore costly, as well as being liable to be injured by exposure to the naked fire; and to avoid these inconveniences, M. Leon Droux, of Paris, has designed and patented an apparatus, which has already been introduced in several large works. A full-size apparatus of this kind is also exhibited by M. Droux in Class 51 of the French section. This apparatus consists of a large cylindrical copper vessel, 2 ft. in diameter, into which the fatty matters to be decomposed are introduced by means of a funnel and cock. The copper cylinder is capable of resisting safely an internal pressure of about 213 lbs. per square inch; and it is enclosed for the greater part of its length in the outer iron cylinder, capable of resisting a similar pressure to the copper cylinder. The lower part of this iron cylinder, which contains water, is placed in a furnace, and the cylinder thus serves as a water-bath for the inner vessel, and as a boiler for producing the steam required for effecting the decomposition of the fatty matters. The pressure also being the same within the iron as within the copper cylinder, all the lower part of the latter is relieved from strain. Altogether the apparatus is neatly arranged, and seems well adapted for its intended purpose.

Iron Workshops.—Self-acting Tools.—The Exhibition of 1862 showed the first signs of a new era in tool-making having commenced on the Continent. The erection of railway repairing-shops all over the Continent kept up a constant supply of orders; cheap labour, short distances for carriage, and in many cases a high protective tariff, assisted the first steps in this direction. The tool-makers sold their tools to others, but imported English machines for their own use, until they had a plant sufficient to enable them to arrive at the same degree of accuracy in workmanship which used to be the distinguishing feature of machines made in the best workshops of England. The ability and industry of their workmen has always been, and is still, an important point in favour of continental manufacturers. The result could not fail to be what it has been, viz., equality in every respect except the cost of production and selling price of the articles.   So much for the history of German tool-making during the last few years. In France the course followed was similar, only modified by the peculiar method of action, which seems to suit the national character in that country, the Government taking the subject in hand.   The Conservatoire des Arts et Métiers, that excellent national institution, began the collection of designs of English machines, and placed them at the disposal of everybody who was willing to take copies from them. Designs are furnished, according to requirement, to the other national schools and educational establishments throughout the country, and are obtained by the constructors and draughtsmen of the different engineering establishments, practically free of all expense.   In some establishments in France the productions stand upon a level with the best tools produced in any country.

Locomotives.—The English show is very small; but it has the honour of contributing by much the finest locomotive to the Exhibition—a model of strength, symmetry, compactness, and simplicity.   Go in among the French and German works and see what you find there. You find scores of engines, some of them extremely well made.   One is a locomotive ordered for our Great Eastern Railway.   Forty such locomotives have been ordered; 15 of which have been approved, and this is the sixteenth.   Such a fact as this may not be generally known, and it is important.   When we go into Wurtemberg we find another fine locomotive ordered for England, at least for one of the East India railways.   An order has been given for no less than 20 similar ones. We still provide the designs for the locomotives, but we have to go abroad to get them manufactured, where the labour market is cheaper. A question naturally arises—Will not the designs follow the manufacture?   Here you go to Creuzot for 40 engines as good as any that can be made in England, and cheaper; or you go to Esslingen for 20 of similar merit.   What will the end of this process be?   Pride ourselves on our designs as we may, how long are we likely to retain superiority of design if we cannot also maintain superiority of manufacture?

Nail-cutting Machine.—The Wickersham Nail Company, of Boston, in the United States, exhibit, in the Paris Exhibition, a machine for cutting nails out of plate iron, and which, from the rapidity with which it works, and the excellency of its performance, justly deserves the notice which it attracts.   The nails are cut out perfect, with heads and pointed ends, all made in the one operation of stamping-out.   The machine in the Exhibition is fitted with cutters arranged for making nails $2\frac{7}{16}$ in. long, and which are cut from a plate of iron $\frac{3}{32}$ in. thick. Eight nails fall out fully made with each revolution of the driving-shaft, and at the time we saw it the machine was making about 120 strokes per minute—that is to say, it was manufacturing nails at the rate of nearly 1,000 per minute, apparently with the greatest ease.   In the manufacture of smaller nails the rate of produce would, of course, be much more rapid.

Screw-making Machine.—There is a machine of extremely ingenious

z

construction, for making small screws. The machine completes a screw by a set of several operations effected in succession by different tools. These tools are fixed in a revolving holder, which turns after each operation has been completed, and presents to the article operated upon a new cutter, which completes another part of the work. The turning of the tool-holder is made self-acting, the workman having only to move a lever forward and backward for each special operation, each set of six operations completing the work to be performed by the machine upon one article.

Wood-Working Machinery.—Mr. Markert's machine-made doors and windows occupy a large frontage in the Austrian machinery gallery, amongst the objects of civil engineering and architecture. The collection contains doors of the simplest kind, with specimens of flooring and windows to match; and commencing with these it rises through a considerable number of gradations, all represented by full-size specimens, up to the most highly ornamented doors, windows, and "parquet" floorings, such as made at these works for the internal outfit of some of the finest houses and palaces on the Continent. We have before this noticed the extraordinary depth and boldness of Mr. Markert's mouldings, which is sometimes very difficult to produce by machinery, but which gives to these productions a particular architectural beauty.

MISCELLANEOUS.

A machine for door-hinges, and capable of producing sixty complete hinges per minute, was shown in the French department. These hinges are made of two sheets of brass, formed in narrow strips, and coiled up each in a large roll, which unwinds itself while working, and of one coil of wire, which supplies the central pin of the hinge. The strips of brass are punched out in the places where they are intended to interlock each other, and are afterwards doubled round the piece of wire, which is cut to the proper length. Ultimately the holes for the wood screws or nails to pass through are punched and countersunk, and the finished hinge thrown out of the machine.

Machine for Cutting Wood for Lucifer Matches, in a perfectly cylindrical form and of equal lengths, was exhibited in the French department. Matches of cylindrical form have been hitherto produced almost exclusively by hand, particularly in Austria and in some parts of Germany. In those countries a very considerable trade in such matches has been carried on for a very long time, and they are considerably exported to other countries, including England and America. The action of the machine is very rapid, throwing out a small bundle of matches at each stroke, and the quality of the article produced is superior to that made by hand.

Enamelled Iron-ware for Domestic Use.—In the Austrian department some interesting articles were exhibited. There is a copper plate enamelled with a glass-like substance, which is neither affected by moderate heat nor by acids, and does not crack when the plate is beat in either direction to a very sharp curve. The more ready and very successful application made of it hitherto, however, is the production of enamelled iron ware for domestic use and for chemical laboratories.

The enamel is absolutely free from lead, which most frequently forms an element of similar enamels, and is of course injurious. The vessels covered with this material can be exposed to all kinds of rough treatment, and even scratched over with a knife, without destroying the glassy surface, nor does the latter suffer from exposure to heat.

Agricultural Machines.—Messrs. Fowler have a magnificent display of steam-ploughing machinery, whilst Messrs. Howard, of Bedford, have a model of their system in the Annexe, their full-sized tackle being at Billancourt.

The general exhibition is very similar to the one in London in 1862, and the only new objects of importance which we have discovered there are exhibited by Messrs. Ransomes and Sims, of Ipswich, and Messrs. Howard and Co. The latter firm show a new self-acting sheaf-delivery reaper, in which the mechanical apparatus for working the rake which delivers the sheaves is constructed by means of two wheels and a crank, thus avoiding all the complicated gear which has hitherto been found necessary in order to obtain the combined lateral and rotary motion to the rake. Ransomes and Sims also exhibit an improved steam threshing-machine, with apparatus for bruising and chopping the straw so as to fit it for use as fodder for animals. The grain is threshed in the same way as in Ransomes and Sims' ordinary machines. After leaving the shakers, it is passed into a box containing two rollers fitted with knives and revolving at great speed ; one of these rollers cuts the straw into proper lengths, whilst the other bruises it, and, by the rapidity of its motion, and the peculiar shape of the cutters, it acts upon the straw in such a way that it leaves the machine thoroughly softened and prepared for the food of cattle. The chopped straw is blown by a cleverly contrived elevator to any height required.

Superiority in invention does not settle that of superiority in the means of production. It does not much matter where an invention comes from, for generally all the world may adopt and work it. The great test of superiority is cheapness and excellence of production. The French, Belgians, and Germans can now spin cotton, make iron and machinery, and build ships and bridges, as well, and nearly (if not quite) as cheaply as we can. They can supply their own wants better than by buying from us, and they can compete with us in foreign markets, and even in some things now and then place their goods in our own markets. They have plenty of iron and plenty of coal for a very large production, and they can import raw materials from America, Russia, or the East, as cheaply, of course, as we can. They have learnt all our trades, and, although we for a long time refused to give them credit for it, they are really excellent workmen. It is useless to boast longer of our supremacy ; for this has now but a doubtful, if indeed it has any real, existence ; yet, at the same time, we are certain that our Continental rivals have not obtained what we understand as the supremacy, nor will they ever obtain it. They can never become such extensive producers as to supply all the markets of the world, independent of ourselves. But that the French especially have made wonderful progress is shown by official and even startling statistics.

India in the Paris Exhibition, 1867.—Those who interested themselves in the Exhibition of 1862 will certainly notice with regret the absence of any signs of engineering progress in India in the present one.   On the former occasion were to be seen specimens of iron manufacture by the East India Iron Company, who are, however, now content to be represented merely by samples of the native iron ores.   In raw materials India is somewhat better represented, and there are several specimens of iron ore from different parts of the empire exhibited, and in one or two instances specimens of iron and steel prepared from them, and also specimens of charcoal from Indore, Burmah, and elsewhere. We failed to discover any petroleum, which however has now been found in several parts of India.   The products of the country have, perhaps, never been so extensively exhibited before, but the omission to include amongst the articles anything that would exhibit the engineering progress of the country has been to exclude those very branches of industry by which, in a greater degree than any other, India must look, if ever, to be raised from her present position to one of almost fabulous wealth and prosperity, to which her immense natural, but hitherto undeveloped, resources might yet one day raise her.

INDUSTRIAL COLLEGES.*

(Extracts from a Lecture delivered before the London Society of Arts,
by Dr. Lyon Playfair, C.B., F.R.S.)

It is well to inquire, in what we are so deficient, and what is the
reason of this deficiency. Assuredly it does not consist in the absence
of public philanthropy or want of private zeal for education, but
chiefly rests in that education being utterly unsuited to the wants of
the age. In the thirteenth and fourteenth centuries classical learning
was, after its revival, highly esteemed : and its language became the
common medium for expression in all nations. A thorough acquaint-
ance with it was an absolute necessity to any one with pretensions to
learning. It had a glorious literature, one as fresh as when it grew on
the rich soils of Rome and Greece. Its truths were eternal, and were
received by us in their traditional mythology, as Bacon beautifully
says, like " the breath and purer spirit of the earliest knowledge floating
to us in tones made musical by Grecian flutes." And why was that
bewitching literature made the groundwork of our educational systems?
Does it not show that literature, like art, may have a standard excellence ; and that we are content to imitate where we cannot surpass.
If the main object of life were to fabricate literati, I would not dispute
the wisdom of making classics the groundwork of our education. They
are not utterly dead, but, like the dry bones of the valley, they may
come together, and have breathed into them the breath of life. In the
world there is a constant system of regeneration. Theories exist for a
time, but, like the phœnix, are destroyed, and rise yet more glorious
from their ashes. Animals die, and by their decay pass into the
atmosphere, whence vegetables derive their nutriment, and thus death
becomes the source of life. But in all this there is no incongruity.
A phœnix does not from its ashes produce an eagle, but a phœnix as
before. The dry bones of dead literature may vivify into new forms
of literary life. Classical literature and exact science are, however,
wholly antithetic. If classical literature be sufficient to construct
your spinning-jennies and bleach your cottons, your system of instruction is right ; but if you are to be braced and your sinews strengthened
for a hard struggle of Industry, is it wise that you should devour
poetry, while your competitors eat that which forms the muscles and
gives vigour to the sinews ? With such different trainings, who in the
end will win the race? Science has not, like Literature and Art, a
standard of excellence. It is as infinite as the wisdom of God, from
whom it emanates. All ordinary powers decrease as you depart from

* If such a lecture be necessary in a country like England, it is ten thousand times
more so in India, where all the arts and manufactures have been at a stand-still for
hundreds of years.

the centre; but the power of knowledge augments the farther it is removed from the human source from which it was transmitted. God has given to man much mental gratification in trying to understand and apply to human uses His laws. How can we as a nation expect to carry on those manufactures by our sons of Industry, when we do not teach them the nature of the principles involved in their successful prosecution?

All the aspirations of youth are towards Science, especially that depending on observation, but we quench the God-born flame by "freezing drenches of scholastic lore." You know of the nations that have towered high in the world, and the lives of men who have saved whole empires from oblivion. What more will you ever learn? Yet the dismal change is ordained, and then, thin meagre Latin (the same for everybody), with small shreds and patches of Greek, is thrown, like a pauper's pall, over all your early lore; instead of sweet knowledge, vile, monkish, doggerel grammars and graduses, dictionaries and lexicons, and horrible odds and ends of dead languages, are given you for your portion, and down you fall from Roman story to a three-inch scrap of "*Scriptores Romani*"—from Greek poetry, down, down, to the cold rations of "*Poetæ Græci*," cut up by commentators and served out by schoolmasters. Is this horrible quenching of all our youthful innate love of God's truth the education for the youth of a nation depending for its country's progress on their development? How is it possible that dead Literature can be the parent of living Science and of active Industry?

It would ill become me, or any one, to speak disparagingly of the wisdom to be derived from a study of ancient authors, or to deny the immense importance of a knowledge of classical literature to education generally; nor would I like to see that education confined to stern realities, divested of the graces and poetry of polite literature. But I do, at the same time, vehemently protest against the exhaustion of all our youthful years by a mere classical tuition, especially in the case of that large class of the community who, by their exertions in industry, have confided to them, in a great degree, the prosperity of their country. As I do not think the teaching of classical literature as practised in our schools to be worthy of the name of education, neither do I apply that title to the communication of scientific knowledge alone,—and you will observe that I have always spoken of it by the term "instruction." I am propounding no scheme of education, but strongly insisting that instruction in Science should form an important part of the education of our youth. You may, and I hope will, soon raise an industrial university; but this should have its pupils ready trained before it adopts them. Now, it must from itself act downwards, instead of working from the school upwards. Until our schools accept as a living faith that a study of God's works is more fitted to increase the resources of the nation than a study of the amours of Jupiter or of Venus, our industrial colleges will make no material headway against those of the Continent. In Paris, we find a Central College of Arts and Manufactures, into which the students enter at an average age of nineteen years, already well trained in the elements of Science, and going there to be taught how to use these elements for industrial application. Three hundred of the best youth of France are annually receiving at this college the most elaborate education; and the best proof of its practical value is the great demand among manufacturers for its pupils, a diploma from it being equivalent to assured success in life. Can you wonder at the progress making by France in industry, when she pours every year an hundred and fifty of these highly-educated manufacturers into her provinces? A similar education to this is going on in almost all parts of Europe; but in England only one such institution exists. We have our University and King's College, it is true, and they are productive of much good; and similar colleges exist in Scotland and Ireland; but their instruction in Science terminates just where the industrial colleges of the Continent begin. In fact, the latter would be supplementary and a great support to the former. Government, acting on its own perception of right, in its first national recognition of these truths, now happily dawning on England, has established a School of Mines; and the experience of this has shown that it is much appreciated, although it labours under the disadvantage of the want of a preliminary education in its pupils, compelling its professors, in its commencement, to be more elementary in their instruction than is well compatible with the proper objects of such a school. Now, while I urge the impolicy of a mere classical instruction to the youth of this country with all the expression which I can give to a matured conviction, do not suppose that I would wish to put all our youth in one Procrustean bed. I again allege, that it is the present system which follows this singular love of uniformity, and clips or extends the dimension of each youth to one common standard. It is against this very confined system that I protest. It is truly lamentable that Oxford and Cambridge so little encourage the Sciences; for, until the colleges throw open their widest portals to these, the schools in the country, deriving their life from them, will do little to reform the present vices of a limited and exclusive education.

In this country we are, in many respects, remarkably unchangeable. Three professions, the Church, the Law, and Medicine, were supposed, some centuries since, to represent learning, and, with a wonderful blindness, they are still accepted as all-sufficient. Industry, to which this country owes her success among nations, has never been raised to the rank of a profession. For her sons there are no honours, no recognised or social position. Her native dignity, if tacitly understood, has never formally been acknowledged. Science, which has raised her to this eminence, is equally unrecognised in position or honours, and, from her very nature, cannot attain the wealth which in Industry solaces the absence of social position. This restriction of learned honours to three recognised professions has a lamentable effect both on the progress of Science and of Industry. Its consequence is, that each profession becomes glutted with ambitious aspirants, who, finding a greater supply than demand, sink into subordinate positions, becoming soured and disappointed, and therefore dangerous to the community. Raise Industry to the rank of a profession,—as it is in other countries,—give to your industrial universities the power of granting degrees involving high social recognition to those who attain them, and you will draw off the excess of those talented men, to whom the Church, the Bar, and Medicine offer only a slender chance of attaining eminence; and by infusing such talent into Industry, depend upon it the effects will soon become apparent. In foreign countries, professions involving social rank and position arise with their requirements; in our nation, we are content with a meagre classification, scarcely sufficient for the middle ages, and not even a reflection of our present wants. These considerations are not mean ones, for, as long as ambition exists in the human mind, their good or bad adjustment will exercise a beneficial or pernicious influence on society. . . . . . Science has been a prime cause of creating for us the inexhaustible wealth of manufactures; and it is by Science that it must be preserved and extended. We are interested as a commercial people—we are interested as a free people. The age of glory of a nation is likewise the age of its security. The same dignified feeling which urges men to gain a dominion over nature will preserve them from the dominion of slavery.

Rapid transition is taking place in Industry; that the raw material, formerly our capital advantage over other nations, is gradually being equalised in price, and made available to all by the improvements in locomotion; and that Industry must in future be supported, not by a competition of local advantages, but by a competition of intellect. All European nations, except England, have recognised this fact; their thinking men have proclaimed it; their governments have adopted it as a principle of state; and every town has now its schools, in which are taught the scientific principles involved in manufactures, while each metropolis rejoices in an industrial university, teaching how to use the alphabet of Science in reading manufactures aright. . . . . . Our manufacturers were justly astonished at seeing most of the foreign countries rapidly approaching and sometimes excelling us in manufactures, our own by hereditary and traditional right. Though certainly very superior in our common cutlery, we could not claim decided superiority in that applied to surgical instruments, and were beaten in some kind of edge-tools. Neither our swords nor our guns were left with an unquestioned victory. In our plate-glass, my own opinion—and I am sure that of many others—is, that if we were not beaten by Belgium, we certainly were by France. In flint-glass, our ancient *prestige* was left very doubtful, and the only important discoveries in this manufacture were not those shown on the English side. Belgium, which has deprived us of so much of our American trade in woollen manufactures, found herself approached by competitors hitherto almost unknown; for Russia had risen to eminence in this branch, and the German woollens did not shame their birthplace. In silversmith work we had introduced a large number of foreign workmen as modellers and designers; but, nevertheless, we met with worthy competitors. In calico-printing and paper-staining our designs looked wonderfully French; whilst our colours, though generally as brilliant in themselves, did not appear to nearly so much advantage, from a want of harmony in their arrangement. In earthenware we were masters, as of old; but in china and in porcelain our general excellence was stoutly denied; although individual excellencies were very apparent. In hardware we maintained our superiority, but were manifestly surprised at the rapid advances making by many other nations. . . . . . It is a grave matter for reflection, whether the Exhibition did not show very clearly and distinctly that the rate of industrial advance of many European nations, even of those who were obviously in our rear, was at a greater rate than our own; and if it were so, as I believe it to have been, it does not require much acumen to perceive that in a long race the fastest sailing ships will win, even though they are for a time behind.

A competition in Industry must, in an advanced stage of civilisation, be a competition of intellect. The influénce of capital may purchase you for a time foreign talent. Our Manchester calico-printers may, and do, keep foreign designers in France at liberal salaries. Our glass-works may, and do, buy foreign science to aid them in their management. Our potteries may, and do, use foreign talent both in management and design. Our silversmiths and diamond-setters may, and do, depend much upon foreign talent in art and foreign skill in execution. But is all this not a suicidal policy, which must have a termination, not for the individual manufacturer, who wisely buys the talent wherever he can get it, but for the nation, which, careless of the education of her sons, sends our capital abroad as a premium to that intellectual progress which, in our present apathy, is our greatest danger? This points to the necessity of the establishment of industrial colleges; but it implies, at the same time, an adaptation of juvenile education to the wants of the age. All this impresses itself upon my mind with a conviction as strong as that the glorious sun sheds its light-giving rays to this naturally dark world of ours. . . . . . Do not let us, by severing Industry from Science, like a tree from its roots, have the unhappiness of seeing our goodly stem wither and perish by a premature decay; but as the tree itself stretches out its arms to heaven to pray for food, let us, in all humility, ask God also to give us that knowledge of His works which will enable us to use them in promoting the comfort and happiness of His creatures.

LIST OF WORKS CONSULTED,

OR FROM WHICH EXTRACTS HAVE BEEN TAKEN, OR OTHERWISE MADE USE OF FOR THE PURPOSES OF THIS WORK.

Manchester Cotton Supply Reporter.
Baines's History of the Cotton Manufacture.
Mann's Cotton Trade of Great Britain.
Arnold's Cotton Famine.
Watt's Lancashire and the Cotton Famine.
Morris's Past and Present of Cotton Machinery.
Warden's Linen Trade.
Onell's Calico Printing and Dyeing.
Muspratt's Chemistry as applied to Arts.
Reed's History of Sugar.
Burgh's Treatise on Sugar Machinery.
Proteaux's Guide to the Manufacture of Paper.
Herring's Paper and Paper-making.
Hopkinson on the Steam-engine.
Bourne's Handbook of the Steam-engine.
Fairbairn's Mills and Millwork.
Fairbairn's Useful Information for Engineers.
Heads and Hands in the World of Labour.
Hunt's Dictionary of Arts and Manufactures.
Lectures on the Results of the Great Exhibition.
Parnell's Applied Chemistry.
Clarke's Exhibited Machinery.
Birmingham and the Hardware Districts.
Scientific Record of the Great Exhibition.
Illustrations of Useful Arts and Manufactures.
Statistical Abstracts presented to Parliament.
Reports of the Juries, Exhibition, 1862.
Board of Trade Returns.
Trade Circulars, &c. &c.

INDEX.

VIRTUE AND CO., PRINTERS, CITY ROAD, LONDON.